AF869954

CHINA WATER

中国水利
CHINA WATER

Ministry of Water Resources of the People's Republic of China

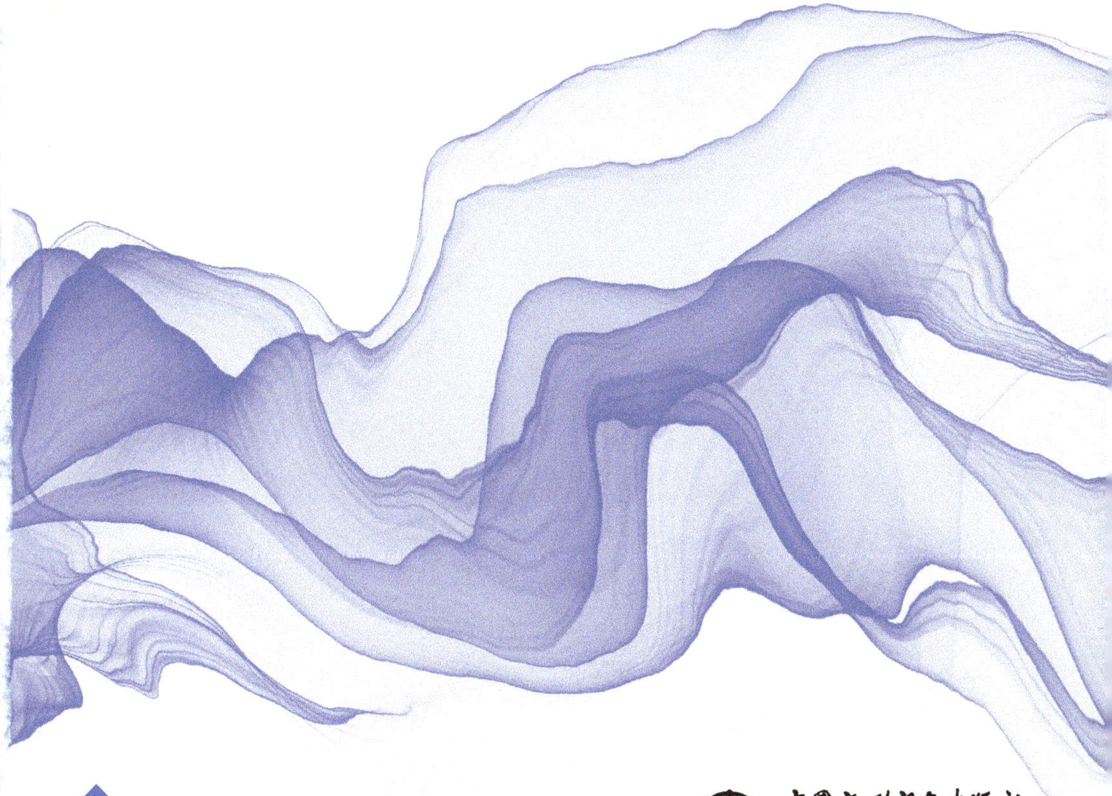

Paths International Ltd

中国水利水电出版社
www.waterpub.com.cn

CONTENTS

Chapter 1
China's Water Management System and Policies

Written by: Ni Peng and Tang Jin

Translated by: Zhong Shumin and Shi Xianglin

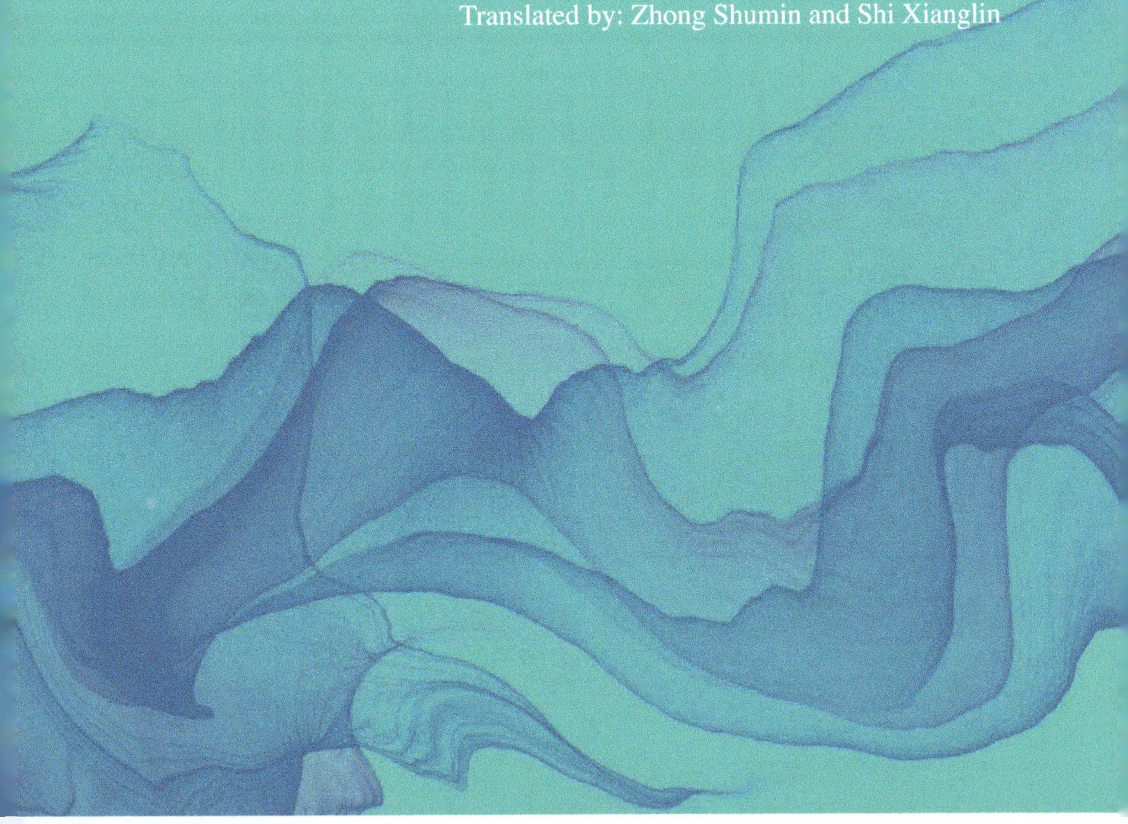

1. Management System

1.1 Water Management System Mechanism

Water resources are owned by the state. The ownership of water resources is exercised by the State Council on behalf of the State. The state operates a management system that combines watershed management with regional administrative management of water resources.

The Ministry of Water Resources, as an important department of the State Council in charge of national water administration, we have continued to uphold the general situation of serving economic and social development, to deepen institutional reforms, and to promote functional transformation. We have established a water management system mechanism that optimizes cooperative efficiency and safeguards water security for economic and social development.

1.1.1 The Water Conservancy Function System is Constantly Improved

With the continuous economic and social development, the strategic importance of water conservancy continues to be elevated. The focus of water administrative management is shifted from water conservancy project construction to water resources management and ecological water protection and to water security for economic and social development. The water conservancy function system continues to be enriched and refined. Especially since the year of 2013, we stick to implementing President Xi Jinping's key spirits as to water management and the water management policy of the new era—"prioritizing water saving, spatial balance, systematic governance, and equal attention to market and government". We have accelerated functional transformation, reformed institutions in 2018, and further strengthened integrated management. Furthermore, we have augmented the unified management of the construction and operation of

major water conservancy projects and water resources allocation and dispatching, reinforced the duties of the most stringent water resources management, water conservation, river and lake management and protection, and taken further steps to focus on our major responsibilities and core businesses. On the basis of maintaining the integrity of the duties of the main body of the Ministry of Water Resources, some duties have been realigned in order to give full play to the professional and management expertise of all relevant departments and thus to establish a coordinate and efficient operational mechanism of the Ministry of Water Resources and related departments taking their respective responsibilities while remaining consistent with one another. The Ministry of Water Resources has set up a functional system with water management as the core, covering water resources development, utilization, conservation, protection, allocation and dispatch.

1.1.2 Water Administrative Agencies are Continuously Improved

The Ministry of Water Resources has undergone many institutional reforms, especially the institutional reform in 2018, in which the former Executive Office of Three Gorges Project Construction Committee of the State Council, Office of the South-to-North Water Diversion Project Commission of the State Council were merged into the Ministry of Water Resources. Improvements in the organizations and institutions of the Ministry of Water Resources have been remarkable. At present, there are 22 divisions within the Ministry of Water Resources: the General Office, the Department of Planning and Programming, the Department of Policy, Law and Regulation, the Department of Finance, the Department of Human resources, the Department of Water Resources Management, National Office of Water Conservation, the Department of Water Project Construction Management, the Department of Water Project Operation Management, the Department of River and Lake Management, the Department of Soil and Water Conservation, the Department of Rural Water and Hydropower, the Department of Water Project Resettlement, the Department of Supervision, the Department of Flood and Drought Disaster Prevention, the Department of Hydrology, the Department of Three Gorges Project Management, the Department of South-to-North Water Diversion Project Management, the Department of Water Diversion Management, the Department of International

Cooperation, Science and Technology, etc. Additionally, the state implements a water resources management system that combines river basin management with regional administrative management. The Ministry of Water Resources has set up river basin management institutions in seven river basins-Yangtze River, Yellow River, Huaihe River, Haihe River, Pearl River, Songhuajiang-liaohe River and Taihu Lake, exercising water administrative responsibilities according to the law within the scope of jurisdiction. Besides, river management agencies that extend to the grassroots level have been established in the Yellow River Basin, Huaihe River Basin, and Haihe River Basin. After 70 years of development, the "1+7+N" water administrative structure of the Ministry of Water Resources, the river basin management organizations and the river management agencies at all levels guarantee effective implementation of water-related administrative responsibilities.

1.1.3 Supporting Institutions Continue to be Strengthened

Currently, there are 28 directly-affiliated institutions to support the Ministry of Water Resources. Their scope of business includes providing comprehensive technical support and guarantee for all water conservancy business areas concerning hydrology and water resources monitoring and forecasting, water conservancy planning and design, water conservancy science and technology research and development and application, water conservancy publicity and education, water conservancy policy research, drinking water safety in rural areas, water conservancy project construction and management, water resources management, water conservation, river and lake management protection, soil and water conservation, etc.

1.2 Main Functions of the Ministry of Water Resources

1.2.1 Ensure water resources are rationally developed and utilized. Formulate water resources development strategies, plans and policies; provide draft legislation and promulgate water administrative rules and regulations; make integrated river basin management plans and flood control plans for major rivers and lakes.

1.2.2 Coordinate and secure domestic, industrial and ecological water uses. Organize the implementation of the most stringent water management system; undertake integrated water resources management and supervision; formulate long-term supply and demand planning and water allocation plans for national and interregional waters and supervise their implementation; be responsible for water resource dispatching of important river basins, regions and major water transfer projects; organize and undertake survey and assessment of water resources, as well as survey of hydropower resources in accordance with relevant provisions; allocate water among major river basins, regions and key water diversion projects; implement the system of water permit, the charging system of water resource fees and systems of water resources assessment and flood assessment. Provide guidance for the water supply work of water industry and towns.

1.2.3 Originate relevant systems for water conservancy construction according to regulations and organize implementation. Make proposals for total investment in water projects with fixed assets and plans of national fiscal funds; review and verify fixed assets investment projects within the national plan and the scale of a yearly plan in accordance with the authorization granted by the State Council. Put forward and organize the implementation of the Central Government's water project investment plan.

1.2.4 Take charge of water resource protection. Organize and draft water resource protection plans; provide guidance on the protection of drinking water

sources, exploitation and utilization of groundwater as well as management and protection of groundwater resources. Organize and guide the comprehensive remediation of groundwater over-exploitation areas.

1.2.5 Take charge of water conservation. Draft water-saving policies; organize water conservation planning and supervise its implementation; develop relevant standards. Organize the implementation of management systems such as total water use control; guide and promote the building of water-saving society.

1.2.6 Provide guidance for hydrological affairs. Take charge of monitoring of water resources, construction and management of national hydrological station network; undertake monitoring of water quantity and quality of rivers, lakes, reservoirs and underground water; release hydrological data and water resources information, forecasting and national water resources bulletin. Carry out the investigation and evaluation of water resources and hydropower resources and the monitoring and early warning of water resources carrying capacity according to regulations.

1.2.7 Provide guidance to the management, protection and comprehensive utilization of water infrastructures, water bodies and shorelines. Guide the management, development and protection of major rivers, lakes, estuaries and coast beaches; give directions to the ecological protection, restoration, the ecological flow and water management, and the connection of rivers and lakes.

1.2.8 Guide and supervise the construction and operation management of water projects. Organize and implement the construction and management of controllable backbone or key water projects that cross the boundary of regions and watersheds; initiate and coordinate the implementation of relevant policies and measures for the operation of the Three Gorges Project, the operation of the South-to-North Water Diversion Project and the construction of follow-up projects; guide and supervise the safe operation of projects; organize related jobs for engineering acceptance; oversee and guide the construction of local supporting projects.

1.2.9 Be responsible for control of soil and water losses. Work out soil and water conservation plans and supervise their practice; enforce comprehensive measures for erosion prevention, monitoring and forecasting, and periodical notification; take charge of the supervision and management of soil and water conservation in construction projects; provide guidance for the implementation of national key water and soil conservation projects.

1.2.10 Guide rural water conservancy work. Organize the construction and transformation of large and medium-sized irrigation and drainage projects; provide guidance on the construction and management of projects for safe drinking water supply and water-saving irrigation; coordinate water development activities in range areas, and establish a socialized rural water service system; offer directions to the exploitation of hydropower resources in rural areas, small hydropower transformation and rural electrification of hydropower.

1.2.11 Guide the management of resettlement work for water projects. Work out policies related to water project resettlement and supervise their practice; organize the implementation of systems concerning acceptance, supervision and evaluation of water project resettlement; guide and supervise the implementation of the follow-up support policies for reservoir migrants; streamline and supervise the support work of the resettlement of the Three Gorges Project and the South-to-North Water Diversion Project; coordinate and promote partnership assistance and other work.

1.2.12 Be responsible for investigating illegal cases violating water laws, mediate and arbitrate inter-sector and inter-province water disputes, and provide guidance on the enforcement and execution of water-related laws and regulations. Take charge of the safe operation of the water conservancy industry in accordance with relevant laws; organize and guide the safety supervision of reservoirs, hydropower dams and rural hydropower stations; guide the supervision and management of the water conservancy construction market; organize the implementation of supervision of water conservancy project construction.

1.2.13 Carry out work related to water science and technology and foreign affairs involving water issues by undertaking quality supervision, draft and promulgate the technical standards, specifications and codes of the water sector and supervise their practice. Handle foreign affairs concerning trans-boundary rivers.

1.2.14 Take charge of satisfying the relevant requirements of comprehensive disaster prevention and mitigation planning; organize the preparation of flood and drought disaster prevention planning and protection standards and guide their implementation. Undertake flood and drought monitoring and early warning, arrange the preparation of flood prevention and drought relief dispatching and emergency water dispatching schemes for major rivers and lakes and key water projects; submit the scheme according to the correct procedure and put it into action; provide technical guidance on emergency rescue of water hazards; undertake the important water project dispatching work during the typhoon defense.

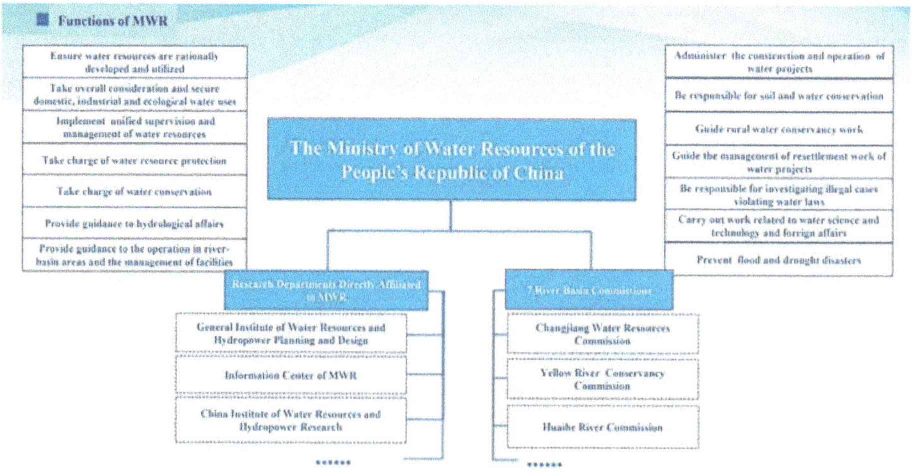

1.2.15 Complete other tasks assigned by the Central Committee of the Communist Party of China and the State Council of the People's Republic of China.

1.2.16 Transform the functions. The Ministry of Water Resources should earnestly strengthen the rational utilization, optimal allocation and conservation of water resources. Adhere to the priority of water saving; shift from increasing supply to paying more attention to demand management; strictly control the total amount of water and improve water efficiency. Adhere to the priority of protection, strengthen the management and protection of water resources, water bodies and water conservancy projects, and maintain the health and beauty of rivers and lakes. Adhere to the priority of overall planning and taking all factors into consideration, ensure rational use of water and sustainable exploitation of water resources, and safeguard water security for the economic and social development.

2. Policies and Legislation

2.1 Fundamental Condition

China's water conservancy legislation extends from complying with the initial problems of radical recovery from water disasters and building water conservancy to effectively solve the further problems of water shortage, water ecological damage, and water environmental pollution, ensuring national water safety. Especially since the reform and opening up, the construction of water conservancy policies and regulations has been pushed forward in an all-round way. A large number of water conservancy policies and regulations have been promulgated one after another. Thus, the system of water laws has been gradually improved, which can be divided into three stages.

The First Stage (1978 - 1987)

In order to establish a legal system for water control and management by law as soon as possible, the Ministry of Water Resources began to draft the water law of the People's Republic of China in April 1978 and initiated legislative work on soil and water conservation as well as water source protection. By the mid-1980s, many important achievements had been made. In 1984, the standing committee of the National People's Congress promulgated the *Law on Prevention* and Control of *Water Pollution*; in 1982 and 1985, the State Council released *Soil and Water Conservation Regulation* and *Water Engineering Regulation of Water Fee Calculation, Collection and Management*. Later, a number of regulations and normative documents were issued by the Ministry of Water Resources, such as *Management Rules for River Embankment Projects,*

Interim Regulations on Irrigation Area Management, Management Regulations on Water Resources and Hydropower Projects, etc. Nearly 50 local water regulations were introduced, covering the management and protection of water resources, water project management, water fee calculation management, irrigation and water conservancy projects, small hydropower development, rural drinking water projects, land acquisition of water conservancy projects and resettlement compensation, etc. The laws and regulations were available in some areas of water management. During this period, the water legislation was mainly embodied in some single laws, but there was no overall plan.

The Second Stage (1988 - 2001)

On January 21, 1988, the first basic law regulating water-related activities - *Water Law of the People's Republic of China* - was promulgated, marking an era of water resources management by law. In 1989, the Ministry of Water Resources formulated the *Overall Plan of Water Law and Regulation System* and adopted different measures to implement it. Water conservancy legislation has gradually been standardized and the legislative process has been greatly accelerated. The *Soil and Water Conservation Law* was announced in 1991 and the *Flood Control Law* was announced in 1997.

In order to implement various systems, governments at all levels and water administrative departments have stepped up the construction of supporting systems. Many administrative regulations and rules, such as *Regulations on River Management, Measures for the Implementation of Water License Systems* and *Relevant Provisions on the Management of Construction Projects within the Scope of River Management*, have been promulgated and implemented successively. In 1997, the State Council issued *the Interim Measures for the Administration of Raising and Using Fund for Water Conservancy Construction* and the *Industrial Policies for Water Conservancy*. Overall progress has also been made in the construction of local water regulations. More than 700 local water

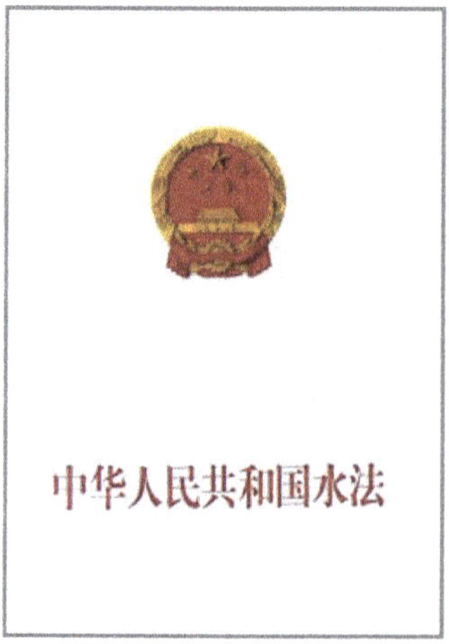

Water Law of the People's Republic of China, the principle law for the regulation of water-related activities, which was promulgated and put into effect on January 21,1988

regulations and regulatory documents have been promulgated, and the water legislative system is becoming more and more complete. The proclamation and implementation of the above water conservancy policies and regulations have had a significant and far-reaching impact on the development of water conservancy.

The Third Stage (2002-now)

In the 21st century, China's construction of the law system has been advanced in depth. Speeding up the construction of ruling the country by law has become one of the focuses of national work. On August 29, 2002, the *Water Law* was revised and passed by the 29th Session of the Standing Committee of the Ninth National People's Congress and came into force on October 1, 2002. The new *Water Law* legalized and institutionalized the state's water control policies in the new era, adhered to the strategy of sustainable development, established the water control concept of harmonious coexistence between man and nature, strengthened the integrated management of water resources, and highlighted the allocation,

Control Indices of "Three Red Lines"

Control index of "Three Red Lines"(y)	Water use quantity of national total (billion m^3)	Water consumption for industrial added value (m^3) per 10 000 yuan RMB	Effective utilization coefficient of farmland irrigation water	Water quality compliance rate in key water function zones of rivers and lakes
2015	635	decrease by 30% from 2010 level	>0.53	60%
2020	670	<65*	>0.55	80%
2030	700	<40*	>0.6	95%
* Calculated by the constant price of 2000				

Implementing the Most Stringent Water Resource Management System

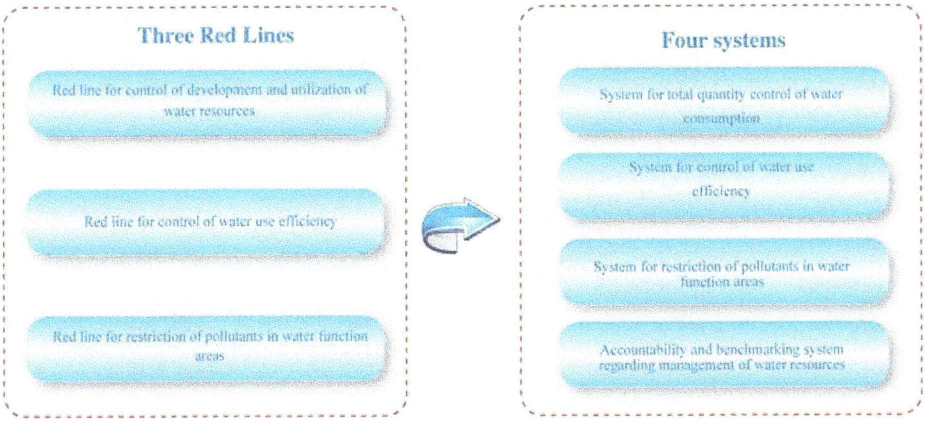

conservation, management and protection of water resources.

In order to implement the new *Water Law*, the Ministry of Water Resources revised the *Overall Plan of Water Regulation System* in 2006 and 2013, highlighting the institution construction of allocation, conservation, management and protection of water resources.

Currently, China has established a relatively complete system of water laws and regulations with the *Water Law* as the core, including 4 laws, 24 administrative regulations and 53 regulations of the Ministry of Water Resources, covering all fields of water conservancy work. Under the framework of the *Water Law*, *Law on Prevention and Control of Water Pollution*, *Regulations on Water Permission and Collection of Water Resources Fees* and *Regulations on Water Supply Management of the South-to-North Water Diversion Project*, the water resources management system has been ranged from planning, allocation, dispatching, water permit and paid use to conservation and protection. Under the framework of the *Water Law*, *Regulation on River Management* and *Regulation on Sand Production in the Yangtze River*, river-management system has included scope delineation of river course, river-related activities control, sand production management as well as river protection. Under the framework of the *Flood Control Law*, the *Flood Control Regulations* and *Drought Relief Regulations*, flood control and flood fighting, drought relief and disaster reduction, flood storage and detention area management, and protection of flood control engineering facilities have been involved in. Represented by the *Regulations on Compensation for Land Expropriation and Resettlement for Large and Medium-sized Water Conservancy and Hydropower Projects* and the *Regulations on Farmland Water Conservancy*, the water conservancy project construction and management system includes land expropriation and resettlement, quality and safety, operation and management, etc. The *Soil and Water Conservation Law* covers the system of prevention and control of soil erosion. The water regulation system is based on the Constitution. It not only adapts to the national conditions and water system and reflects the characteristics of water conservancy, but also strengthens cohesion and coordination with other laws and regulations. At the same time, local governments at all levels have continuously strengthened the construction of the water regulation system and achieved remarkable results. A total of more than 900 local regulations and government regulations have been issued, laying a solid institutional foundation for the comprehensive management of water according to the law.

On April 2, 2019, the First Meeting of China-EU Dialogue Mechanism for Water Policies was Held in Beijing. E Jingping, Minister of the Ministry of Water Resources, Delivered a Keynote Speech.

2.2 Developmental Strategy and Main Measures

Entering a new era, water conservancy legislation adheres to the guidance of President Xi Jinping's new concept and new strategy of comprehensively governing the country according to law. The implementation outline for the construction of a government under the rule of law, closely centers on the implementation of the major development strategies of the Party and the country. For water conservancy, it focuses on the policy of "prioritizing water saving, spatial balance, systematic governance, and equal attention to market and government", and the general tone of water conservancy reform and development, "Accelerating the construction of water conservancy projects and strengthen the supervision of water conservancy industry" was accordingly formed. With the evolution of major social contradictions, major water control contradictions and changes in the development situation and tasks of water conservancy reform, we will continue to deepen the reform and promote the construction of the rule of law. We will continue to use the law to standardize all aspects of water conservancy reform and development, improve the water law more suitable for China's national conditions and water system, and provide legal safeguard for strengthening water management and ensuring water safety.

2.2.1 Strengthening the Top-level Design of Water Conservancy Legislation

Based on the legislative priorities of the Party and the state and the actual needs of water conservancy reform and development, new legislation, amendments, abolishment and interpretation of the law have been carried out simultaneously. An overall plan has been made for laws and regulations. The top-level design of legislation has been strengthened. The *Overall Planning for the Water Law System* has been revised and implemented to clarify the legislative tasks, priorities and project layout, accelerating water conservancy legislation, and gradually forming a systematic and coordinated water law system.

2.2.2 Strengthening the Legislation in Key Fields

In order to improve the quality and efficiency of legislation, preliminary research of the comprehensive and strategic water regulation system has been actively carried out. Focusing on the implementation of the major developmental strategy of the Party and the state, the general principle of water conservancy reform and development, as well as the strengthening of social management and public services, we have accelerated the legislative process of water laws and regulations. These include Yangtze River protection, water conservation, groundwater management, river and lake management, sand mining in river courses, water regulation in river basins, rural drinking water safety, etc. We have improved the laws and regulations on flood and drought disaster prevention, soil and water conservation, rural water conservancy and hydropower, hydrological management, water conservancy project construction as well as its management and protection. So the requirements of comprehensively implementing the governing of the country by law will be put into all aspects of the process of water conservancy reform and development.

2.2.3 Meeting the New Demand Based on the General Tone

We adhere to the problem-oriented principle, focus on the implementation of the general tone of water sector, highlight key fields and critical joints, improve the systems of relevant rules and regulations, follow standard specification and implementation methods, standardize construction in the early stage, improve the scheduling and safety of water conservancy projects, specify the regulatory scope, methods, and responsibilities, ensure that the weak factors in the water conservancy project are improved and strong supervision is carried in water sector. All is done by laws and rules.

2.2.4 Improving the Quality and Efficiency of Legislation

We make laws in a scientific, democratic and legislative way, improve the system of collection and demonstration, review and investigation, and evaluation of legislative projects. We furthermore carry out legislative research, major system design and special research on difficult issues to improve the quality of legislation and enhance the pertinence, systematicness, operability and effectiveness of legislation. We will strengthen legislative coordination, expand public participation and properly handle the demands of all parties. We will improve the post-legislative evaluation system and make overall plans for new legislation, amendments, abolishment and interpretation of water laws and regulations.

The Hukou Waterfall which is located in the Yellow River on the Shanxi-Shaanxi border

Chapter 2
Water Resources Management and Protection

Written by: Wang Haiyang
Translated by: Shi Xianglin

1. An Overview of Water Resource

Total volume of water resources in China stands at 2.8 trillion m^3, ranking 6th in the world, whereas water resources per capita in China is only 2100 m^3, 28% of the world average.

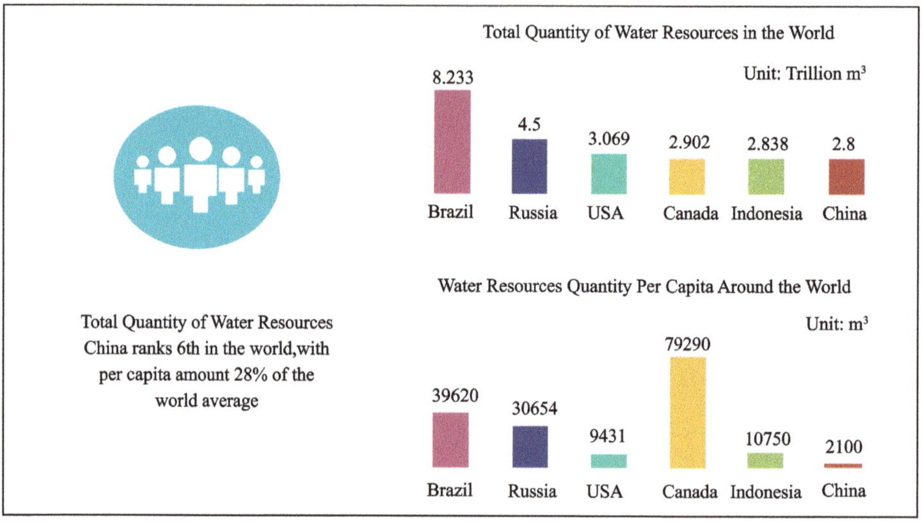

Water resources are unevenly distributed in space, and don't match the layout of land resources and production forces. North China takes up 64% and 46% of the total national land area and population respectively but only 19% of national water resources.

Water resources are unevenly distributed in time. Precipitation and river runoff in China are highly concentrated within a year, but vary substantially between years. For most regions, precipitation in 4 consecutive months accounts for 60%~80% of the annual total. In North China, precipitation in the wettest years could be 3~6 times of that in the driest years; and for river runoff, the corresponding difference could be more than 10 times.

2. Major Achievements

The Chinese government takes resource conservation and environmental

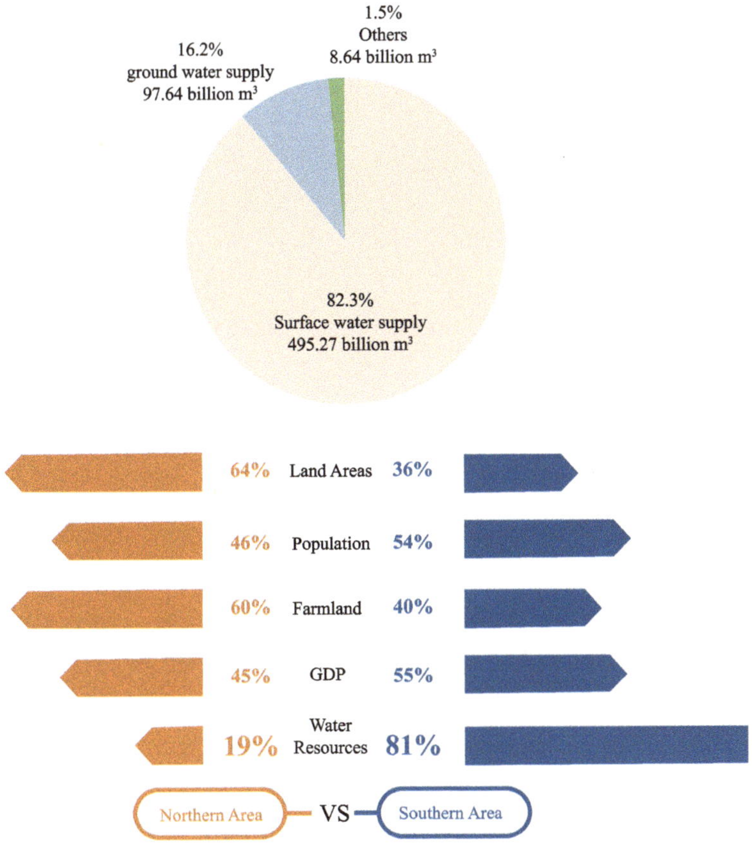

Unmatched Layout between Water Resources and Economic Development

protection as a fundamental policy, and implements the most stringent water-resource management system. Focusing on the total quantity and strengthening the control of water consumed, the government promotes strong supervision of water resources to accelerate the construction of an ecological civilization and water-saving society. Since 2012, China's total annual water consumption has remained at about 610 billion m^3, with GDP increasing from 54 trillion yuan to 90 trillion yuan. The rapid economic development of society has been supported by the micro-growth of total water consumption.

2.1 Development and Utilization of Water Resources

Since 1949, water development and utilization in China has undergone 3 phases:

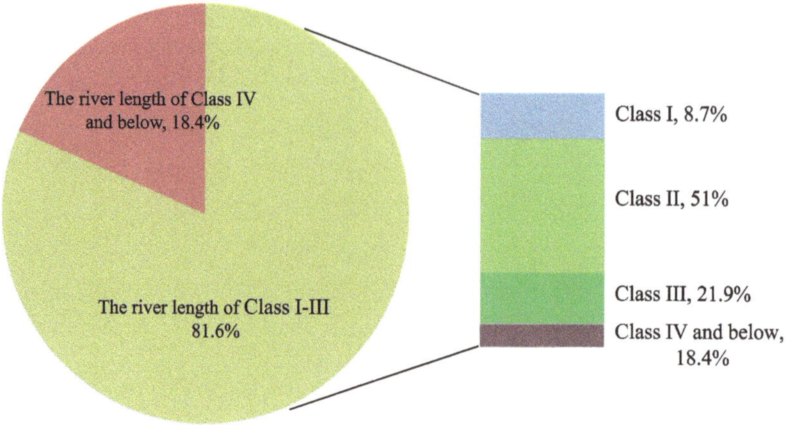

The Annual Water Quality Type Ratio of Chinese Rivers in 2018

single-purpose development, multi-purpose development and sustainable development. In 2018, China recorded 601.55 billion m^3 of both water supply and consumption, 432 m^3 comprehensive water consumption per capita, 66.8 m^3 water consumption per RMB 10,000 GDP (at current year price), 41.3 m^3 water consumption per RMB 10,000 industrial added value (at current year price), 225 L/d domestic urban water consumption per capita, and 89 L/d domestic rural water consumption per capita.

In 2018, the water quality of rivers with a coverage of about 262,000 km in China was evaluated. The river length of class I-III and the river length of class IV and below respectively constituted 81.6% and 18.4%. Compared with 2017, the ratio of river length of class I-III increased by 1.0%, while the ratio of poor river length of class V decreased by 1.3%.

2.2 Allocation of Water Resources

The country has approved water allocation plans of 41 inter-provincial river basins, defining the rights and interests in inter-regional areas. The overall water dispatch of the Yellow River, Heihe River, Tarim River, Hanjiang River and other river basins has been implemented. The main stream of the Yellow River has flowed continuously for 20 consecutive years. The Dong Juyan Lake in the lower reaches of Heihe River hasn't dried up for 15 consecutive years. The unified water dispatch of the first phase of the East Middle Route of the South-to-North Water Diversion Project has been strengthened. The total amount of water transferred by the South-to-North Water Diversion Project has exceeded 27 billion m^3, with a direct beneficiary population of about 120 million. Water resource augmentation in new urban areas, industrial parks and major industry layouts has been executed. Meanwhile, the country has implemented water-based city planning and water-based production planning with strict water permit management. This basically solved the issue of water permit in large-scale and key medium-sized irrigation areas in the centralized water-supply regions in most northern provinces.

2.3 Protection of Water Resources

The government has formulated and promulgated the *National Water Resources Protection Plan* and revised the *Measures for the Supervision and Administration of Water Function Zones*. It implemented the division of national water function zones and strengthened the supervision of water function zones and sewage outlets into rivers. At present, the monitoring coverage rate of important water function zones in the country is 95%. The safety of drinking water sources has greatly improved by approving the announcement of 618 important drinking water sources nationwide and organizing the construction, inspection and evaluation of safety and security standards. The construction of 105 Water Ecological Civilized Cities in the country as a pilot project has been further promoted. Furthermore, the government has supported local implementation of a river-lake-reservoir water system connection project, focusing on water ecological restoration to effectively improve the condition of rivers and lakes. The government carried out the study on determining the ecological flow of rivers and lakes, and organized the formulation of the work plan and policy for ensuring the ecological flow of 21 key rivers and lakes.

2.4 Comprehensive Treatment of Over-exploitation of Groundwater.

Over-exploitation of groundwater has been included in the focus of water resources, water ecology and water environmental protection to increase efforts in governance. The *National Groundwater Utilization and Protection Plan*

Restored Wetlands in the Yellow River Estuary

was issued to determine national groundwater protection and management goals, tasks and measures. Efforts have been accelerated to comprehensively treat over-exploitation of groundwater in key areas. Six provinces and cities including Beijing, Tianjin, Hebei, Henan, Shandong and Jiangsu have taken full advantage of the South-to-North Water Diversion Project to replace the groundwater in urban areas where water is needed. The annual exploitation of groundwater in urban areas has been reduced by about 1.9 billion m^3. The comprehensive treatment of over-exploitation of groundwater focused on North China, including Hebei, Shandong, Shanxi, and Henan, where the annual amount of exploited groundwater in rural areas has been reduced by at 2.1 billion m^3. Also, the government has implemented a pilot for the comprehensive treatment of over-exploitation of groundwater in rivers and lakes in North China. They also used the Middle Route of the South-to-North Water Diversion Project and the reservoirs in Hebei Province to recharge groundwater in 3 rivers of Hebei Province. Up to the end of August 2019, a total of 1.32 billion m^3 has been recharged with remarkable results in improving the ecology of rivers and lakes. The *Action Plan for Comprehensive Control of Over-exploitation of Groundwater in North China* has been issued and implemented.

2.5 Reform of Water Resources in Key Areas

The government has organized and completed seven water-rights pilot projects in Ningxia, Inner Mongolia, Guangdong, Henan, Gansu, Jiangxi and Hubei,

provinces made substantial exploration of the right to use water resources, transaction circulation and system construction. A variety of water-rights transaction models has been formed, which include water-rights among inter-basin, inter-region, upper and lower reaches of river basins, inter-industry and inter-user. The country has successively launched pilot projects for tax reform of water resources in 10 provinces and areas including Hebei, Beijing, Tianjin and Shanxi, formulated and issued the *Opinions on Reform of the Paid Use System of Water Resources*, and made deployment and requirements on water-resource fee collection and management, pilot projects of tax reform in water resources, and water-rights confirmation and transactions.

2.6 Basic capability of Water Resources Management

The government has accelerated the construction of the national water-resource monitoring system and completed the online monitoring of important water users, quantity and quality of river sections in important provincial boundary. Additional routine monitoring of 521 important sources of surface water and drinking water throughout the country has also taken place. If water quality exceeds the set standard, there will be a pre-warning treatment. A national water resources information management system covering deployment at the central, provincial and municipal levels and five-level application has been basically completed. A relatively reasonable national-level groundwater monitoring network has been constructed, covering 31 provinces, autonomous regions and municipalities in the country and the main plain areas of Xinjiang Production and Construction Corps.

3. Future Development Strategies and Main Measures

With economic and social development, the problem of Chinese water resource safety is more prominent. President Xi Jinping, standing at the height of the country's overall development with the aim of improving national water safety and promoting the development of an ecological civilization, puts forward the idea of "prioritizing water saving, spatial balance, systematic governance, and equal attention to market and government". In September 2019, President Xi Jinping proposed at the symposium on ecological protection and high-quality development of the Yellow River Basin that water resources should be considered as the most rigid constraint. With this in mind, we reasonably plan the development of population, cities and industry, and resolutely curb the unreasonable demand for water. China will continue adhering to the basic national policy of saving resources and protecting the environment, focusing on two major work goals of "reasonable water distribution and water-use control", to solve the key issues of water shortage and water-related ecological damage. Meanwhile, China will take various measures to achieve this, such as strictly control water intake and ecological flow, refine the management of serious water-related ecological problems and thoroughly implement the most stringent water resource management system. Other measures include continuously improving the capability and level of water resource supervision, constantly adjusting people's behaviour through strong supervision, correcting people's misconduct, restraining unreasonable water consumption demands, and promoting high-quality development.

3.1 Work Goals

The government will control the total quantity of water consumed, strictly abide by the limits of water resource development and utilization, prevent the over-exploitation of water resources, and protect water resources and water ecology. Meanwhile, the government will establish a fundamental safeguard system for the ecological flow of rivers and lakes and effectively implement control measures. Nowadays, over 1000 km^2 of trans-provincial river basin flow distribution has been completed and water-intake project checking and registration of seven basins have been comprehensively completed. Meanwhile, it is essential to implement key monitoring of water resource basins, regions, rivers and lakes with over-development, to reinforce the control of total exploitation of groundwater and the water level, and to reinforce governance in over-exploitation groundwater areas, so as to relieve the situation of groundwater over-exploitation nationwide. The government will also establish the water resources monitoring system in important rivers, realize overall coverage of important river sections, and focus on realizing online monitoring of important water intake and drinking water sources. They will further deepen water resource tax reform and the reform of the water-rights system, and improve the supervision capability of water resources.

3.2 Main Measures

3.2.1 Confirm and Control Ecological Flow

The government should accelerate the understanding of the ecological flow of rivers and lakes, formulate a list of important rivers and lakes for ecological flow protection, implement the list-based management of ecological flow protection, and determine the protection goals for ecological flow. For this, we should fully consider water resource conditions, engineering regulation conditions and water demand for life, ecology and production, then we classify, formulate and implement the differentiated and targeted control measures. It is necessary to implement river and lake water allocation plans, reinforce regional water permit plan management, implement unified dispatching of river basin, and strengthen the supervision of operation and scheduling for wading projects on rivers to implement ecological flow control goals. Beyond this, we should accelerate the construction of a monitoring network for ecological flow in important river sections, carry out a pre-warning system and risk control, establish a responsibility system for ecological flow and allocate responsibilities to different departments, include the implementing performance of controlling ecological flow to the most stringent assessment of water resources management system, and intensify the accountability of supervision and assessment to put ecological flow management into practice.

3.2.2 Promote the Water Distribution of Water Basins

The government should accelerate the water distribution of trans-provincial rivers and further allocate the sharing of them with approved distribution into cities and counties. The government should intensify supervision of the implementation of the approved river distribution plans, formulate and implement the water dispatching plans in river basins and the annual water dispatching plans, unify the dispatching of the water in river basins and extremely important water projects, while the total quantity of regional water use and water release in river sections should be controlled, water permit supervision should be intensified, and various control goals and measures of water distribution should be implemented.

3.2.3 Strictly Implement Water Permit Supervision Management

The government should comprehensively carry out the checking and registration of water withdrawal projects to grasp their current situation relating to water withdrawal projects (or facilities) and the management of water resource development and utilization in various river basins. Considering the existing problem, a standing book for controlling the total water use in river basins

Ecological Improvement by Integrated Water Resources Management in the Lower Reach of the Tarim River

should be established to implement dynamic calculating of the total volume of water withdrawal and consumption, while differentiated regional plans should be organized and formulated. In areas where the total volume of water consumed reaches or exceeds the quota there should be limited approval of water permit.

A Pilot Project of Water-saving Society Development, Zhangye, Gansu Province

The planning of water resources argumentation should be promoted. The control indexes for water resource in different areas should be defined to implement the policy of water-based production, which means what we can do is decided by the total water we can use. We should deepen the reform of the approval system for water permit. The "all in one network" and the electronic permit dependent on government integration service platform should both be implemented, providing a more simplified approval system, stronger supervision, and better service. The supervision of water withdrawals from water resources should be reinforced. Inspection of key water resources should be intensified. Water withdrawal behavior should be standardized in water users. Responsibilities of supervision and management of water resources development and utilization will be pushed into implementation.

3.2.4 Reinforce Groundwater Protection and Over-exploitation Governance

We should carry out double control of the total volume of groundwater exploitation and water level, mark out the forbidden areas of groundwater exploitation, and manage the paid use of groundwater. Groundwater investigation, evaluation and monitoring should be perfected, while technical support of groundwater management is provided. What's more, it is essential to enlarge the groundwater supervision of main over-exploitation groundwater areas in the Northeast, North China and the Northwest to accurately grasp the dynamic changes of regional groundwater. Focusing on the governance of the Beijing-Tianjin-Hebei region, we will accelerate the governance of groundwater over-exploitation in North China according to the comprehensive management idea of "one decrease and one increase". Through diversified channels, ecological water compensation of rivers and lakes in North China can be enlarged, ecosystems of rivers and lakes will be restored, underground water will be recharged."Water-based" production in Xi Liaohe basin, Inner Mongolia, should be completed and the excessive development and utilization of water resources, intensification of cutoff of river courses, and the over-exploitation of groundwater should be systematically handled.

3.2.5 Reinforce Drinking Water Resource Protection and Connect Rivers and Lakes

We should reinforce the main responsibilities of local governments, intensify monitoring of drinking water source quantity and quality, and promote the sharing of such data with relevant departments. Based on the construction project of capability monitoring of national water resources, it is necessary to implement dynamic monitoring of water sources, and enhance the ability to be alert on the risk and control it to improve drinking water safety. We will promote the establishment of a joint prevention and control mechanism for water pollution incidents in the upper and lower reaches of inter-provincial river basins. Local areas will be instructed to intensify standby water source construction to further guarantee their water supply safety. We should guide local authorities to intensify early-warning through research and judgement, joint monitoring and information reporting, scientifically dispatch water and implement pollution prevention and control measures. Focusing on the river and lake systems choked with silt, we will gradually restore the integrity of the lake and river systems and improve or restore the hydraulic connection of the river, lake and reservoir systems. We will drive the formation of standards and technical systems for the ecological protection and restoration of rivers and lakes, and carry out condition assessments of rivers and lakes.

3.2.6 Strengthen the Basic Work of Monitoring, Measuring and Gathering

Statistics of Water Resources

According to the construction project of monitoring capability of national water resources, we should speed up the development and construction of water resource monitoring system, formulate and issue policies about reinforcing management of measurement for water withdrawal and consumption, and perfect the technical specification and standard system for such measurement. We should improve the equipment for water resource measuring and fully implement the application and transformation of intelligent measuring instruments for water-use in key units with water withdrawal and consumption. Furthermore, we should establish and improve the list of basic units in the national water-use statistical survey, improve the information system of the direct reporting platform for water consumption statistics, and improve the standing book for water withdrawal in regions and units. The statistics and checking system of water use, the auditing system of statistic data of water use and the accountability system should all be established.

3.2.7 Constantly Deepen the Water Resources Reform in Key Fields

We should comprehensively implement and develop tax reform for water resources, force to standardize the behavior code for water withdrawal and consumption, promote the conservation, protection and optimal allocation of water resources.We should deepen the reform of water permit by streamlining administration and delegating powers, fully implement the electronic water permit, realize "all in one network, information sharing", further optimize the administrative examination and approval procedures, improve the efficiency of examination and approval, improve the ability and level of water permit supervision; study and push up the property rights system of water resource assets, actively and steadily promote the confirmation of water rights, cultivate and develop the water market, and fully develop the role of the market in optimizing the allocation of water resources.

The Namco Lake, Tibet Autonomous Region

Chapter 3
Water Saving

Written by: Zhao Zhixuan
Translated by: Shi Xianglin

1. Status of Water Saving

1.1 Total Quantity Control of Water Consumed

China has paid high attention to water saving for a long time. A slight increase of water consumption, which results from water saving, guaranteed the rapid economic and social development in China.

In 2018, the total quantity of water consumed in China reached 601.55 billion m^3, including 85.99 billion m^3 of water for residential use, 126.16 billion m^3 of water for industrial use, 369.31 billion m^3 of water for agricultural use, and 20.09 billion m^3 of water for ecological environments, which respectively accounted for 14.3%, 21.0%, 61.4% and 3.3% of the total quantity of water consumed.

During 1997-2018, the national water consumption increased to 601.55 billion m^3 from 556.603 billion m^3, an increase of 44.947 billion m^3. In the same period, water for residential use in China increased to 85.99 billion m^3 from 52.515 billion m^3, an increase of 9.33 billion m^3. Water for industrial use increased to 126.16 billion m^3 from 112.116 billion m^3, an increase of 14.044 billion m^3. However, water for agricultural use decreased to 369.31 billion m^3 from 391.972 billion m^3, a decrease of 22.662 billion m^3.

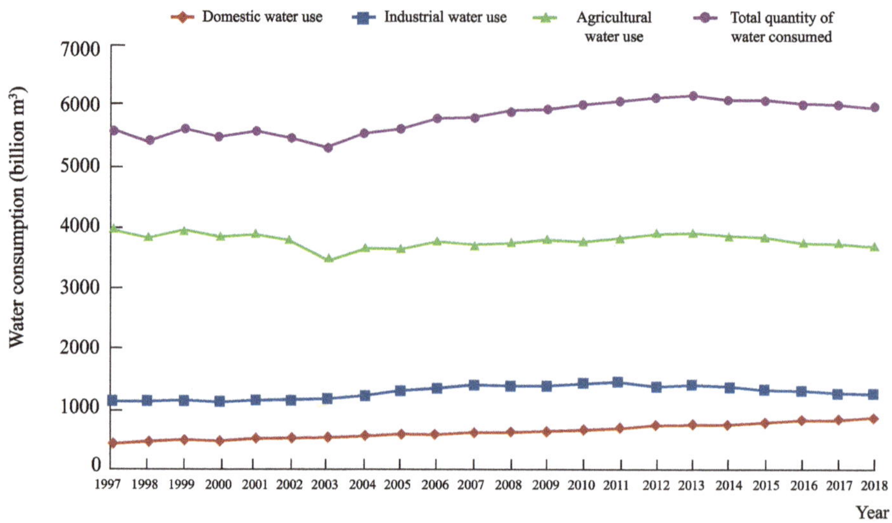

The National Water Consumption Diagram during1997-2018

1.2 Water-use Efficiency Control

In 2018, the national comprehensive water consumption per capita was 432m^3. The water consumption of Gross Domestic Production (GDP) (at the current year's price) per ten thousand yuan was 66.8 m^3. The average water consumption of actual irrigation per hectares in arable land was 5475 m^3. The effective utilization coefficient of farm irrigation water was 0.554. The water consumption of the industrial added value (at the current year's price) per ten thousand yuan was 41.3 m^3. The urban water for residential use per capita (including municipal water use) was 225 L/d and the rural water for residential use per capita was 89 L/d. Since 1997, water-use efficiency has obviously improved, while the water consumption of GDP per ten thousand yuan and the water consumption of the industrial added value per ten thousand yuan were both significantly reduced. Furthermore, the average water consumption of actual irrigation per hectare of arable land had a slow downtrend. By comparing 2018 and 1997, the average water consumption of actual irrigation per hectare of arable land dropped to 5475 m^3 from 7380 m^3. Calculated at the comparable price in 2000, the water consumption of GDP per ten thousand yuan and the water consumption of the industrial added value for ten thousand yuan respectively decreased by 82% and 83% within 21 years.

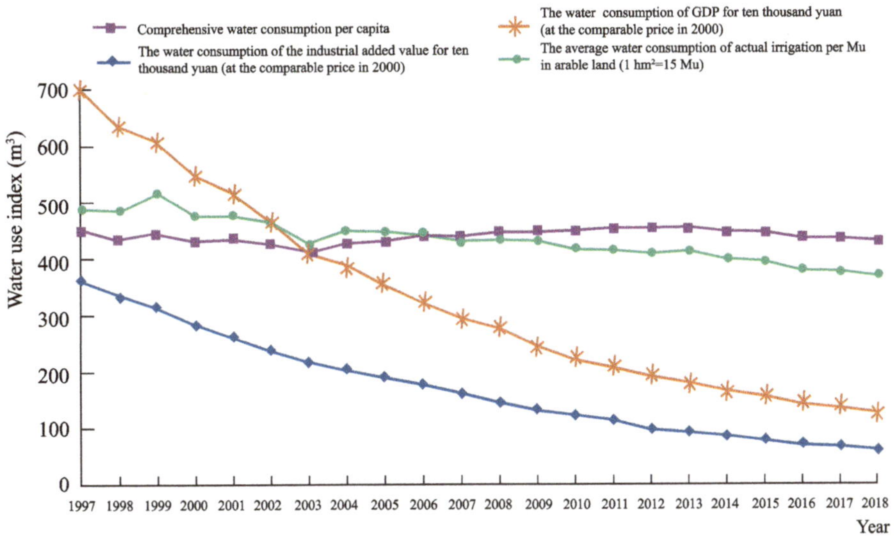

The National Water Use Index Diagram during1997-2018

Domestic Water-Saving: 36 Colleges and Universities in Tianjin Fully Popularizes IC card for Water-saving Intelligent Control System, with Nearly One Million Tons of Water Saved Yearly

2. Main Achievements

Due to an increased population and less water in China, as well as the uneven spatial-temporal distribution of water resources, the disparity between supply and demand is relatively prominent. There is no doubt that the issue of water resources has a great influence on Chinese ecological civil-construction and sustainable economic development in society. Currently, the key is saving water and therefore emphasis should be put on water-saving policy. A related focus is pollution control, similar to saving energy and reducing emissions. The emphasis put on saving water ensures the protection of our ecosystems. Protecting water sources is akin to protecting the homeland and caring about water resources and saving water has become a fashionable and self-conscious action of the whole society. In April 2019, the National Development and Reform Commission and Ministry of Water Resources issued the *National Water-saving Action Plan*, symbolizing that water saving has become a national commitment. Recently, China's water saving has achieved a significant result.

2.1 Water-saving System Construction Being Gradually Enhanced

A series of policy documents and guidance systems have been successively issued in China. For example, the *Dual Control Action Plan on Total water Consumption and Intensity, Water-saving Society Construction in the 13th Five-Year Plan,* the *Action Plan for National Water-saving*, as well as planning policy for water use and water efficiency label are put into practice. We promote the comprehensive reform of the price for agricultural water use, establish the graduated water price for urban residential use as well as an over-quota progressive fare increase system and implement water-saving contract management. Together with the National Water-saving Action Plan and its division schemes in 2019, the water-saving policies and institutional system have been further improved.

2.2 Continuous Progress of Water-saving Technical Transformation

The government reinforces water-saving technical research and development, promotes water-saving facilities, and accelerates water-saving technical transformation. They also utilize these for continuous construction in large and middle-sized irrigation areas, industries with high water consumption, and urban water supply networks. Besides, the government constantly expands unconventional water resources and continuously enhances agricultural, industrial and urban water-use efficiency. Until 2018, the national water-saving irrigation areas have been up to 35.30 million hm^2, which occupied half of the national irrigated farmlands. Among these, the high-efficiency water-saving irrigation areas of sprinkling irrigation and micro-irrigation have reached to 22 million hm^2.

Grain Increase with Water-saving Appears in the Northeast China

2.3 Continuous Breakthroughs in Water-saving Innovation

Grain increase with water-saving appears in the Northeast. Water-saving with high efficiency appears in the northwest. Water-saving with control exploitation of groundwater appears in North China. Water-saving with pollution control appears in the southeast. Water-saving with emission reduction appears in the South. A hundred national-level water-saving pilot programs are in construction, and 65 water-saving counties (districts) have reached a set standard. Around the country, 79,000 water-saving irrigation areas, enterprises, and residential areas have been constructed to develop a good demonstrative leading role.

2.4 Constantly Deepening of Water-saving Public Education

During the "World Water Day" and "China Water Week", the government concentrates on carrying out water-saving publicity activities, develops the brand of publicity activities, such as "Water Saving on the Way" and "Water Saving all over China", promotes the fundamental education of proper water usage, and positively emphasizes the benefits of being close to water, cherishing water and saving water.

3. Developmental Strategies and Main Measures

Nowadays, China is in the decisive stage of comprehensively building "a moderately prosperous society" whilst developing an environmentally friendly construction and developmental model. What's more, now is the crucial stage of implementing the "water-saving priority" policy and breaking down the bottleneck of national water security. In the future, we will stick to implementing the "water-saving priority" policy, promote the completion of various tasks under the guidance of the national water-saving action, drive the transformation of the water-use model from the extensive to the intensive, and promote the sustainable economic development of society based on the proper utilization of water resources.

3.1 Work Goals

When the "moderately prosperous society" is comprehensively built in 2020, the water-saving policies and rules will work with market mechanisms. Besides, the capacity of technical support and methods of management will be improved to increase the rate of saving water. By comparison with 2015, the water consumption of per RMB 10,000 GDP and the water consumption per RMB 10,000 industrial added value will respectively decrease by 23% and 20%. The recycling rate of industrial water-use in a certain scale will be 91%. The effective utilization coefficient of farm irrigation water will be enhanced to 0.55. The leakage rate of national public water supply networks will be controlled within 10%. An advanced system of rules for water consumption system to deal with main crops, industrial products and life services will be constructed. The prefecture-level cities with water-deficiencies will reach the national water-saving standards and water-saving regulations will be issued.

In 2022, water-saving production and lifestyles will be preliminarily established. The water-saving industry will begin to take shape and water-use efficiency will be remarkably improved. The water-saving consciousness of the whole society will be obviously enhanced. The water consumption of GDP per ten thousand yuan and the water consumption of the industrial added value per ten thousand yuan will be respectively dropped by 30% and 28% compared with 2015. The effective utilization coefficient of farm irrigation water will be raised to 0.56. The total quantity of water consumed will be limited to 670 billion m^3. Towns in the Beijing-Tianjin-Hebei region aim for balance between groundwater extraction and supplementation. The amount of unconventional water resource use in cities with water shortages will be improved by an average of 2%. 50% and 30% of county-level (district-level) administrative regions in the north and south respectively will reach the water-saving social standards.

3.2 Main Measures

It is necessary to implement national water-saving action in key areas, and deepen institutional reform through policy promotion and market mechanism innovation.

3.2.1 Strict Control of Total Quantity and Water Requirement Rate

It is essential to construct the double-control index system involved in the provincial, municipal and county-level administrative regions, reinforce water-saving index management, reinforce water permission and water resource assessment, intensify supervision of major water users and special water users, and promote construction of water-saving infrastructures.

3.2.2 Continuously Promoting the Effect of Agricultural Water-saving

It is necessary to accelerate construction of followup and supporting facilities, as well as replacement of outdated facilities. It is equally significant to extensively promote regional high-efficiency water-saving irrigation projects. It is also important to introduce and promote the water-saving technologies of sprinkling irrigation, micro-irrigation, trickle irrigation and water delivery irrigation of low-pressure pipelines. Further measures include assigning plantation on the basis of water yield, developing water-saving systems in fishery and animal husbandry, reinforcing the transformation of rural water facilities for residential use, and promoting the utilization of water-saving facilities.

3.2.3 Greatly Implementing and Promoting Industrial Water-saving and Emission Reduction

It is essential to strongly promote water-saving technologies and skills including high-efficient cooling, washing, recirculating water and the recycling of effluent sewage. It is also necessary to accelerate wastewater treatment, to be reused when the appropriate standards are reached in enterprises with high water-consumption. We should also promote high-quality green upgrades with the emphasis on saving water in existing enterprises and industrial parks, by accelerating facility construction for water saving, circulation, and serial or multi-use between enterprises.

3.2.4 Continuously Promoting Urban Water-saving with Loss Reduction

It is necessary to implement water-saving in urban planning, construction and management, to focus on promoting water-saving transformation in areas with high-leakage networks, and to dramatically reduce leakage loss of water supply networks. In public areas we must improve the installation rate of water-saving facilities and their popularity with urban residents, whilst strictly controlling water use in service industries with high water-consumption. In special industries, such as car washing, golf courses and artificial ski resorts, we should promote the use of recycled water technology.

3.2.5 Deeply Promoting Water-Saving with Open Sources in Major Areas

It is essential to cut down the amount of groundwater mining in over-exploited areas and forbid the utilization of underground water in industry, agriculture and the service industry. Unconventional water use in water-shortage areas should be reinforced and promoted to increase the ratio year by year, and take advantage of seawater in industries and industrial parks with high-consumption water in coastal areas.

3.2.6 Accelerating Leading Role of Technical Innovation

It is necessary to establish technical innovation in water-saving with the in-depth integration of industry-university-research cooperation. This will develop the key technical equipment used for water-saving, intensify the integration of new information technology and expand water-saving technological achievements. The promotion of channels for advanced water-saving technology will facilitate the transformation of the water-saving industry.

3.2.7 Strengthening the Water-saving Policy Reform

It is essential to improve the water-saving standard rating system, execute a strict standard rating, implement comprehensive reform of prices for agricultural water use, promote the graduated water price system for urban residents and an over-

Water-saving Irrigation Brings Good Harvest to Farmers in Panzhihua Mountain Area of Sichuan Province

quota progressive fare increase system for non-residential water use, facilitate water resource tax reform, construct the water-saving statistical investigation and grass-roots water use statistics management system, intensify water-concerned information management, implement the water use report system, and put water use units in industrial and service industry parameters with a certain scale into the monitoring system of major water use units.

3.2.8 Intensifying the Mechanism Innovation in Water-saving Market

It is necessary to promote reform in the water-conservancy market, determine the appropriate use of water resources, reinforce supervision of the water-use trade, and standardize transactions. Also it is necessary to implement proper labelling for water products with high water-saving potentials and wide application, promote water-saving contract management in public institutions, public buildings, industries with high water-consumption and the service industry. Finally, it is necessary to become a leader in water-use products, water-use enterprises, public institutions and water-saving cities.

4. International Cooperation and Exchanges

There is no doubt that the government of China attaches great importance to the international exchanges and cooperation in water-saving fields and has successively developed friendly cooperation with Japan, South Korea, Singapore, Malaysia, Denmark and Israel. Also, the government carries out water-saving projects, including the *Demonstration Project of Sino-Japanese Water-saving Social Construction*, and implements bilateral exchanges and inspection visiting about water-saving with the Ministry of Water Resources in South Korea, Singapore, Malaysia, Denmark and Israel. In the future, through various means, the government will continue to expand and reinforce the exchanges and cooperation with worldwide countries and organizations in water-saving philosophy and strategy, water-saving policy and management, water-saving standards and technology.

Chapter 4
Construction and Operation Management of Water Conservancy Projects

Written by: Wang Shu and Han Han
Translated by: Xiao Rui

1. Overview

Since 1949, the Chinese government has placed water resources administration and utilization in a strategic position of national development and carried out massive water conservancy projects. During the 19th National People's Congress of the CPC, water sector ranks first out of the nine major infrastructure construction missions. Since then, more efforts have been exerted to better accelerate the construction of water conservancy infrastructure.

With consistent efforts made in the past 70 years, China's comprehensive water project system has got into shape, playing a key role in river flood control, farmland irrigation, and urban and rural water supply. The management of project construction is ever-modernizing, accompanied with a significant improvement in construction quality and a safer working-environment at construction sites. Furthermore, the water project system has seen great progress in flood control and drought defense, witnessing a sharp rise of comprehensive benefits for economic and social development.

2.Major Achievements

2.1 Water Conservancy Project Construction

Nowadays, the total amount of reservoirs in China has increased to nearly 100,000, up from 1,200 in 1949. The aggregated water storage capacity has increased from about 20 billion m^3 to about 900 billion m^3. River levees (Level 5 or above for flood prevention cover a combined length of over 300,000 km, enabling China to potentially defend against the greatest flood that has ever occurred in the nation's major rivers or streams since 1949. China has established over 100,000 sluices above designated size, over 95,000 pumping stations, supplying water to 22,000 irrigated zones that covers an area of over 133.33 hm^2. These facilities enable China to guarantee food security for 21% of the world's population,

Taocha Canal Headwork of the Central Route of South-to-North Water Diversion Project

with about 6% of the world's water resources and 9% of the world's arable land. Exemplified by the South-to-North Water Diversion Project, China's water-resource allocation system characterized with the "South-to-North Transfer, East-West Complementary Transfer Strategy". Now the national capacity of water supply has exceeded 860 billion m^3. The water conservancy projects have played a significant role in effective water supply for urban and rural areas, even in medium drought years. Consistent efforts have brought preliminary improvement in 1.2 million km^2 of the land area suffering from soil and water erosion, together with the establishment of 20 million hm^2 of terrace fields and over 60,000 soil-retaining dams. These soil and water conservation measures can reduce soil erosion by 1.8 billion tons annually and enlarge water-retention capacity by 30 billion-plus m^3. In rural areas, over 47,000 hydropower stations have been built, with an installed capacity of 79 million kW, generating a total of 250 billion kW·h of electricity per year. China's water conservancy projects have been ranking among the top of the world both in scale and number. These projects are safe, stable and generating economic and social benefits as planned.

World Bank Loan Project Xiaolangdi Multi-Purpose Hydro Project of the Yellow River

2.1.1 Historic Technological Advances from Imitation to Independent Innovation

Up to date, China has established a relatively comprehensive technological system in water projects. It is home to the world's highest concrete arch dam, roller compacted concrete dam (RCCD), and concrete face rock-fill dam. It also boasts of the most dams that are 200 meters high or above, the worlds' top in installed hydropower capacity and scale of flood discharge structures. China has developed a series of internationally leading technologies, including super-high dam construction, heavy-flow flood discharge, the super-large chamber, the treatment of complicated ecological foundations and super-high slopes.

2.1.2 Modernization of Construction Management for Water Conservancy Projects

Investment in water conservancy construction has gradually shifted from a single channel of budgetary appropriation to multi-channel. This integrates fiscal appropriation, water conservation funding, policy loans, public financing and foreign investment. China has initially established a diverse, multi-level, multi-channel landscape of water conservation development led by government investment and complimented with public investment. Management of water conservancy projects are getting more professional, followed with the full implementation of regulations such

as the legal project person accountability system, tendering and bidding system, construction supervising system and contract management system. With these measures, the market entry and supervision systems are more effective. The credit system is well under construction and various construction methods are in pilot, like agent construction and the general contracting of "design-to-build".

2.1.3 Significant Improvement of Construction Quality

China boasts the lowest rate of dam-break accidents around the world. For decades, it has boasted major-and-super-major-accident-free achievements in water conservancy project construction. 13 hydro-projects, including the Three Gorge Project and the Yellow River Xiaolangdi Project, have won the Milestone Award from the International Commission on Large Dams (ICOLD). Besides, 20-plus projects have been awarded with the China Luban Construction Engineering Prize (National Prime Quality Project); 10 projects with Tien-yow Jeme Civil Engineering Prize; and 100-plus projects with China Dayu Quality Water Conservancy Project Prize.

2.1.4 Progress of Water Conservancy Projects related to Hong Kong, Macao and Taiwan

The Donjiang-Shenzhen Water Supply Project has supplied water to Hong Kong for 50-plus consecutive years and the Fujian-Jinmen Water Supply

Dongjiang-Shenzhen Water Supply Project

Project has started to function. The Datengxia Gorge Water Conservancy Project is well under construction, which will play a significant role in reducing the saltwater tide from downstream, and consequently guaranteeing the security of the water supply in Macao and the Pearl River Delta areas.

2.2 Operation Management

2.2.1 Improvement of Water Conservancy Project Management

China has witnessed much progress in institutional reform of State-owned large and medium-sized water conservancy project management enterprises, resulting in a better separation of project management and maintenance. Institutional reform in small water conservancy project management has also advanced smoothly. This has led to a better clarification of project ownership, improved designation of management and maintenance units and obligations, an increase of funding, and an improved management service.

2.2.2 Solidified Basis for Water Conservancy Project Management

2.2.2.1 The Number of Projects

Carry out water conservancy project census. Up to date, 96,000-plus

Danjiangkou Water Control Project

reservoirs have been registered and documented. China has built a National Reservoirs and Dams Database Management Information System. Other databases, such as the *First Phase of National Large Reservoirs and Dams Safety Monitoring and Supervising Platform*, and *Basic Information Database of Levees and Sluices Projects* are progressing as scheduled.

2.2.2.2 Reservoir and Dam Safety

Put into effect responsibility system for reservoir and dam safety. China strictly implements a responsibility system for reservoir and dam safety, and a system of "Three-person Responsibility Rule" where there is an Administrator, technician, and on-duty inspector for flood control. Every year before the predicted flood season, the media will publicize officials who are responsible for reservoir and dam safety, so as to increase public scrutiny and therefore ensure comprehensive management and responsibility of reservoir and dam safety.

2.2.3 A Fairly Robust Management System on Water Conservancy Project Operation

China has managed to put in place a fairly robust legal system on water conservancy project operation that centers around, among other things, the

Water Law of the People's Republic of China, *Flood Control Regulations*, *Regulations on the Administration of Reservoir and Dam safety*, and *Regulation of the People's Republic of China on the Administration of River Courses*. China has implemented a series of safety systems for administration and technology that cover affairs such as licensing and registration, safety assessment, risk elimination and reinforcement, degrading and decommissioning, dispatching and utilization, repairs and maintenance, monitoring and risk warning, and emergency management and preparation.

2.2.4 Familiarity with Safety Conditions of Water Conservancy Projects

2.2.4.1 Actively Engaging Reservoir and Sluice Safety Condition Assessment Work

China will put into place a reservoir and dam safety assessment warning system, to promptly re-examine the safety issues of those defective reservoirs that need reinforcement or repair, and carry out diagnostic investigations on risky issues of the small-sized reservoirs.

2.2.4.2 Gradual Establishment of Project Safety Monitoring System

Hubei Province has set up a hydrologic information management system of all the small reservoirs in the province. The information system comprises automated flood reporting, SMS alert and real-time image transfer for all 5,200-plus reservoirs. Guangxi Zhuang Autonomous Region has installed monitor terminals at 4,313 reservoirs. Sichuan Province has managed to establish dynamic monitoring and risk alert system for 2,700-plus reservoirs. Ningxia Hui Autonomous Region has initiated construction of a safety monitoring system for the levees of the Yellow River in the region.

2.2.5 Further Reinforcement of Supervising and Examining Project Operation Management

2.2.5.1 Continuous Investigation of the Safe Operation of Water Conservancy Projects in a Secretive Way

In 2018, the Ministry, for the first time, conducted secret investigations of the safe operation of 4,702 small reservoirs. In 2019, the Ministry carried out special investigation of the safe operation of 6,000-plus reservoirs.

2.2.5.2 Inspection of Water Conservancy Project Operation Management

In 2018, the Ministry, on nine separate occasions, sent out a total of 33 inspection teams to investigate 60 reservoirs, 18 sluices and 11 levees located in 25 provinces including Tianjin, Liaoning and Xinjiang Production and Construction Corps. The teams also inspected the administration of the watershed departments and the units afllicted to the Ministry. During the inspection, 575 problems were resolved. Since then, the MWR carried out re-examination of those projects that previously were identified risky and defective. Results show that of those risky reservoirs identified in 2016 and 2017, 84.8% and 65.4% have managed to fix problems or deficiencies respectively.

2.2.5.3 Steady Advancement of Water Conservancy Project Management Assessment

The nation has designed an assessment system to standardize the project management for water conservancy. The measures have effectively facilitated an upgrade of the project management service. By 2019, up to 135 water conservancy project management enterprises have been labeled as "quality management" during the assessment organized by the Ministry of Water Resources.

3. Work Goals

Following the guiding line of goal-oriented and problem-oriented work, we will steadily advance development of major water conservancy projects with effective management and supervision of construction quality. In the meantime, it is our priority to ensure construction safety, prevent and eliminate risks and potential dangers, and generally improve the quality of projects. This is conducive to the healthy development of the water conservancy construction market. It is our prime goal to ensure reservoir construction safety, especially of small ones. Our short-term goal is to conduct a project census and strengthen our foundation of project management. Our medium- and long-term goals include the establishment of a standardized, digitalized and modernized project management system, which lays a solid foundation for the comprehensive development of water conservancy project management.

4. International Cooperation and Exchanges

Water conservancy and hydropower has been one of the first industries in China to "go abroad". Chinese hydroelectric enterprises actively engage in international cooperation. With decades of overseas business and development, these enterprises have gained international reputation in overseas contracting, investment, business and trade in the area of water resources and hydropower. These enterprises are also in command of leading technologies in water resources and hydropower development, operation management, financial services and capital funding. This allows the enterprises to be capable of integrating the whole industrial chain that includes the design, construction and manufacture of major machinery. We have forged diverse cooperation with 100-plus countries and regions in terms of hydropower development, undertaking hydropower projects and river planning in over 60 countries. Doing businesses in over 140 countries internationally, Chinese hydroelectric enterprises own more than 10 million kW of installed capacity rights [①] ,with a total contract amount of US$150 billion, ranking top among all the"go abroad" industries.

In recent years, international cooperation on trans-boundary rivers have been developing smoothly. The Horgos River Friendship Water Diversion Project and the Sumbe River Water Diversion Project between China and Kazakhstan are already in use and generating economic benefits. The Sino-Kazakhstan Horgos River Alamali (Chukulbrack) Joint Mudslide Blocking Dam that has begun construction will, once completed, effectively protect major infrastructure in the lower stream, and ensure the safety of life and property along the river.

The Sino-Kazakhstan Horgos River Friendship Water Diversion Project

① Installed capacity rights refers to the installed capacity according to the proportion of rights. For example, if two companies invest in a power plant with the installed capacity of 1 million kW, and one of them accounts for 60% of shareholder rights, or 60% of all shares, then its installed capacity rights are 600,000 kw (1,000,000 × 60%= 600,000).

Chapter 5
River and Lake Management

Written by: Fu Jian
Translated by: Zhong Shumin

1. Overview

Rivers and lakes carry water resources, provide storage space during floods, and are an important component of the national ecosystem. These functions provide essential support for economic and social development, and play an irreplaceable role in flood control, water supply, power generation, shipping, fisheries, maintaining biodiversity and regulating climate.

China has a large number of rivers and lakes with a long history. A total of 45,203 rivers cover a drainage area of 50 km^2 or more and their combined total length is 1,508,500 km. 2,865 natural lakes occupy a water surface area of 1 km^2 or more. The total surface area of all the lakes is 78,000 km^2. From the south to the north, there are seven major rivers: Pearl River, Yangtze River, Huaihe River, Yellow River, Haihe River, Liaohe River and Songhuajiang River, and five major freshwater lakes: Poyang Lake, Dongting Lake, Taihu Lake, Hongze Lake and Chaohu Lake. Among them, the Yangtze River, with a length of 6,296 km is the third longest river in the world and Poyang Lake is China's largest freshwater lake. At present, China's major cities, infrastructure and grain production areas are mainly distributed along the rivers, namely the flood control zones of seven

major rivers. Their population accounts for about one-third of the country, while their cultivated land accounts for 22%, and the economic aggregate accounts for about half.

With the focus on economic and social development, the impact of human activities on rivers and lakes is gradually expanding. China has established competent authorities for rivers and lakes, improved the management of operations and institutions, refined policies, regulations and planning standards, and strengthened protection. Especially since 2016, the system of river chiefs and lake chiefs has been introduced all over the country. With the main tasks of protecting water resources, preventing water pollution, improving water environments and rehabilitating water ecology, we will gradually build a river and lake management & protection mechanism with clear responsibilities, orderly coordination, strict supervision and forceful protection, and therefore maintain the health of rivers and lakes.

2. Major Achievements

2.1 River Chiefs and Lake Chiefs Grow from "having their names" to "being worthy of their names"

2.1.1 A comprehensive System of River Chiefs and Lake Chiefs has been Established Around the Country

By the end of 2018, 60 local government heads of provincial-level party committees across the nation had been appointed as general river chiefs. More than 300,000 river chiefs and lake chiefs have been hired at provincial, city, county and township levels. Besides, there are over 900,000 village-level river chiefs, lake chiefs, river inspectors and river guards. Thus all the river chiefs and lake chiefs have been assigned their specific names. According to their respective circumstances, local governments have appointed river sergeants and recruited folk river chiefs plus river and lake cleaners. They also developed the role of the volunteer service teams and mobilized the whole society to participate in water management, water control and water conservation.

2.1.2 River Chiefs and Lake Chiefs at All Levels are Fulfilling Their Duties

River chiefs and lake chiefs at all levels inspect, investigate and survey rivers and lakes. They keep informed of the basic situation of rivers and lakes, discover any problems of water erosion, and coordinate and supervise the implementation of renovation tasks. According to statistics, in 2018, river chiefs and lake chiefs at provincial, city, county and township levels inspected rivers or lakes for 7.17 million person-times. Among them, the provincial-level river chiefs and lake chiefs inspected the rivers or the lakes for 1,610 person-times, and the

provincial-level general river chiefs inspected the rivers for 347 person-times.

2.1.3 The Laws, Regulations and the System are Constantly Improving

In various regions, the formation of laws and regulations has been strengthened. Four provinces have promulgated special regional laws and regulations concerning the system of river chiefs and lake chiefs. More than 10 provinces have specified distinctions on the system of river chiefs and lake chiefs in local regulations. These regulations are related to water resource management & protection, river management, sand mining management, lake management & protection, environmental protection, and prevention and control of water pollution. In accordance with the unified requirements, the localities have established a series of systems, including the conference system of river chiefs, the information sharing system, the information reporting system, the work supervision system, the assessment accountability and initiative system, the acceptance system, and so on. Combined with the local introduction of river chiefs inspecting rivers, work supervision and other supporting systems, these systems secure the successful practice of the system of river chiefs and lake chiefs.

2.2 Prominent Problems are Tackled by Iron Fists

2.2.1 Launch a Special Campaign of "Clearing up the Four Disorders" in the Country's Rivers and Lakes

Aimed at the prominent problems in rivers and lakes such as indiscriminate occupation and abuse, indiscriminate mining and over-exploitation, the Ministry of Water Resources has deployed a nationwide campaign of "Clearing up the Four Disorders" in the country's rivers and lakes. Emphasis is put on clearing up and rectifying outstanding problems such as disorderly occupation, mining, heaping and construction. At the same time, the three-year action plan for rural human settlements improvement is combined with the implementation of the strategy of rural revitalization. Focusing on clearing and cleaning, we should speed up "Clearing up the Four Disorders" in rural rivers and lakes and concentrate on solving the complex problems of "dirty disorderly bad".

2.2.2 Start a Special Action for the Protection of the Yangtze River

According to the important instructions of President Xi Jinping's "great protection and no big development", in accordance with the requirements of "unconventional measures, unconventional strength and unconventional effect", we organize three major special rectification actions for the Yangtze River main stream, the sand mining and the solid waste of the Yangtze River economic belt.

Special Rectification Action for Protection and Utilization of the Shoreline on the Yangtze River

Special Rectification Action for the Sand Mining

2.2.3 Carry out Special Actions of River and Lake Cleaning and Renovation in Line with Local Circumstances

Different strategies for rivers and lakes have been devised in different regions, such as organizing and carrying out special campaigns like "limpid rivers", "lucid waters", "clear streams" and "ecological rivers and lakes", promoting the co-governance of water and shore, and implementing systematic management. In some regions, the policy of returning polders and fields to lakes is implemented to restore the ecological space along the shoreline of the river and lake waters. In some regions, rainwater and sewage are separated, water systems are connected, ecological water replenishment and dredging are carried out to eliminate black and smelly water bodies and poor class V water bodies. In some regions, they apply high standards to the construction of "5,000-km green paths" and a safe ecological water system to create clean water and a green bank. In some other regions, river and lake harnessing has even managed to force industrial transformation, the upgrading of the agricultural structure and the transformation of economic development methods, leading to a benign interaction between river and lake management & protection and economic development.

2.3 Reinforce River and Lake Supervision and Inspection

2.3.1 Increase the Number of Unannounced Visits

The Ministry of Water Resources establishes a supervision and inspection mechanism of the system of river chiefs and lake chiefs based on secret visits. In 2018 and 2019, two rounds of secret visits and surveillance were organized, adopting a "four no & two straight" approach and giving full coverage of undercover visits to all districts and cities in 31 provinces (autonomous regions and municipalities). Major problems were highlighted and rectified according to the principle of "one province, one list of responsibilities".

2.3.2 Be Listed to Superintend Major Cases

In response to the prominent problems seriously affecting the health of rivers and lakes, the Ministry of Water Resources, in conjunction with relevant departments, is listed to supervise them. The Ministry will guide and urge the relevant local authorities to seriously investigate and deal with major violations of laws and regulations involving rivers and lakes, which constitutes a powerful deterrent.

2.3.3 Strengthen Social Supervision and Science and Technology Supervision

The supervision e-mail addresses and informants' hot-line telephone numbers have been announced to the public. Many regions have also released their public WeChat accounts, opened 24-hour supervision calls, and created a "hands-on" column to provide complaints and reporting channels for the masses. The public consciously cares for and guards rivers and lakes, and effectively supervises violations of laws and regulations. Remote satellite imaging, unmanned drones and unmanned ships are employed to investigate and verify problems concerning rivers and lakes.

2.4 Consolidate the Basis of River and Lake Management

2.4.1 Speed up the Delineation of the Scope of River and Lake Administration

Accelerate the clarification of border control for rivers and lakes, and the people's governments at or above the county level shall announce the results of the delimitation to the public.

2.4.2 Speed up the Compilation of River and Lake Related Planning

Promote the protection and utilization planning for the shorelines of major rivers and the planning of river sand mining. Improve planning and strict control for the ecology of rivers and lakes.

2.4.3 Adhere to Problem Orientation and Implement Policies in Accordance with Related Rivers and Lakes

Conduct background investigations of rivers and lakes, find out the current real situation of rivers and lakes, identify the main problems, establish one file per river/lake to achieve a good understanding of the whole situation.

2.4.4 Be Dedicated to Building Smart Rivers and Lakes

Establish a national management information platform for the system of river chiefs, develop river/lake supervision apps and supervision information systems, coordinate information sharing related to river and lake management & protection. Many places have implemented the "Internet + the system of river chiefs" action plan, researched and developed a trinity management information system, for the system of river chiefs, that integrates PCs, Mobile clients and WeChat public accounts. Some places set up a Cloud Operational Command Platform for river chiefs, and others set up palm water control circles. All these efforts help to bring about the dynamic supervision of the river chiefs, intelligent management of water, and establish a strict network of rivers and lakes management & protection.

3. Specific Measures

Adopt a holistic approach to conserving our mountains, rivers, forests, farmlands, lakes, and grassland, and harness both rivers and shores. Adhere to local and overall systems, considering both symptoms and root causes, to achieve gradual and long-term convergence. Strive to fight the tough battle of river and lake management, give impetus to the system of river chiefs and lake chiefs to ensure the rivers' peace, people's safety, and citizen's well-being.

3.1 Focus on the "basin" holding water, win the battle of annihilation against the "Four Disorders" in rivers and lakes across the country, reduce stock and set limits to incremental increases. By the end of 2019, the task of centralized cleaning up and renovation will almost be completed, and the natural ecological space of rivers and lakes will be restored.

3.2 Protect the water in the "basin", adhere to the priority of water saving, strictly control water consumption, and strengthen water environmental protection. Increase investment guarantees, systematically deal with issues like "too much water, shortage of water, dirty water and muddy water". Shift from "regulating the chaos" to "treating the disease" and "eradicating the roots", and raise national water security standards.

3.3 Improve the unannounced inspection mechanism, fortify listed supervision over major issues, guide public participation and social supervision, conduct strict rectification and accountability, and urge the river chiefs, lake chiefs and responsible units concerned to perform their duties as well as assume their responsibilities.

3.4 Consolidate the basis of river and lake management & protection, promote the improvement of the rule of law in the management & protection of rivers and lakes, speed up the delineation of the scope of river and lake administration, promote the protection and utilization planning of the shorelines of major rivers and the planning of river sand mining, draw a map of rivers and lakes and build smart rivers and lakes.

Chapter 6
Rural Water Supply

Written by: Hu Meng
Translated by: Shi Xianglin

1. Overview

1.1 Construction of Rural Water Supply Project

Limited by natural, economic and social conditions, the water supply foundation in rural China was weak with poor supply facilities and the water source got polluted easily before the 1980s. Many Chinese farmers in that period had no access to water supply facilities, and instead took water from rivers, streams and ponds directly. They had to fetch water from long distances by carriage or carry water in buckets during drought seasons, which badly affected their livelihood and brought risks to their health, production activities and socio-economic development as well.

The Chinese government has always attached great importance to rural water supply. It has successively implemented projects to relieve the difficulties of rural drinking water, ensure and upgrade the safety of rural drinking water. By the end of 2018, the state had arranged a total investment of 422.2 billion yuan, of which 203.8 billion yuan had been invested by the central government. Combined with the original foundation, a relatively complete rural water supply project system has been formed, with a water supply service capacity of 940 million people, basically covering all rural areas. The centralized water supply rate in rural areas has reached 86%, the penetration rate of tap water has reached 81%. The water supply guarantee rate and water quality compliance rate have significantly increased, and rural water supply guarantee conditions have reached the leading level in developing countries.

1.2 Evaluation Standards for Rural Drinking-water safety

In 2004, the Ministry of Water Resources and the Ministry of Health jointly issued the *Indicators and Standards for Rural Drinking-water Safety Assessment*. In March 2018, the Chinese Hydraulic Engineering Society issued the *Guidelines for Rural Drinking-water Safety Assessment* (T/CHES 18-2018)(Hereinafter referred to as Guidelines). In August 2018, the Ministry of Water Resources, the Poverty Alleviation Office of the State Council and the National Health Commission jointly issued the *Notice on Resolutely Winning the Battle for Poverty Alleviation through Safe Drinking Water in Rural Areas* (Water and Agriculture (2018) No.188). It clearly points out that all localities can directly use the Guidelines or, according to the evaluation criteria and in combination with the actual situation of the province, formulate detailed rules for the evaluation of rural drinking water safety. These can be used as the basis for accurate identification of rural drinking-water safety, formulation of solutions and acceptance of standards for poverty alleviation in all provinces.

The Guidelines clearly stipulate that rural drinking water safety means that rural residents can quickly obtain sufficient drinking water, and long-term drinking does not affect people's health. Drinking water safety consists of 4 conditions: water quantity, water quality, water use convenience and water supply guarantee rate. Drinking water can be evaluated as safe only if all 4 conditions reach the specified standard. As long as one item does not reach the standard or the basic standard, it cannot be evaluated as safe or largely safe.

Indicators and standards for Rural Drinking Water Safety Assessment

	Up to standard	Basically up to standard
Quantity	water-riched areas, quantity≥60L/(cap · d) ; water-deficient areas, quantity≥40L/(cap · d)	water-riched areas, quantity≥35L/(cap · d) ; water-deficient areas; quantity ≥20L/(cap · d)
Quality	Water users of water supply projects with a population of more than 10,000 conform to the provisions of *Sanitary Standard for Drinking Water* (GB5749) ; Water users of water supply projects with a population of less than 10,000 and decentralized water supply projects conform to provisions on the quality of rural water supply in GB5749.	The incidence trend of intestinal infectious diseases among the local population is stable and there is no sudden outbreak, the total number of colonies in microbial indexes and disinfectant indexes may not be included in the evaluation indexes. For water users of the decentralized water supply project, there is no visible impurities, no color and odor in water, and there is no adverse reactions after long-term drinking.
Level of convenience	The round trip time for water intake by manpower in water-supply-to-household areas (including residential areas or courtyard) shall not exceed 10 min or the horizontal distance for water intake shall not exceed 400 m and the vertical distance shall not exceed 40 m.	The round trip time for water intake by manpower shall not exceed 20 min, or the horizontal distance for water intake shall not exceed 800 m and the vertical distance shall not exceed 80 m. In pastoral areas, the round trip time of water intake by simple transportation can be used for evaluation.
Guarantee rate	≥95%	≥90%, and<95%

2.Major Achievements and Challenges

2.1 Major Achievements

Terraced Fields at Dengjiaban, Guangze County, Fujian Province

The implementation of the rural drinking water project has effectively solved drinking water problems such as insufficient water quantity and poor water intake, greatly improved water quality, reduced incidence of waterborne diseases, and improved the health level of rural residents. After tap water has reached the farmers, it ended the history of rural residents carrying water from afar and ensured safe drinking water for elderly women and children left behind in rural areas. The rural labor forces were also liberated and the income of rural residents increased. Environmental protection of rural

Beneficiary Population Based on Rural Drinking Water Safety Consolidation and Promotion in Different Regions during 2016-2018

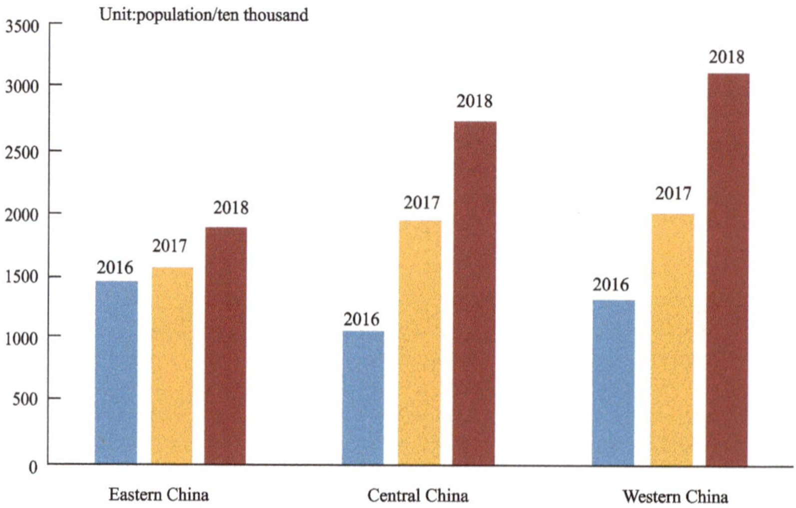

Beneficiary Population Based on Rural Drinking Water Safety Consolidation and Promotion in Rural Areas during 2016-2018

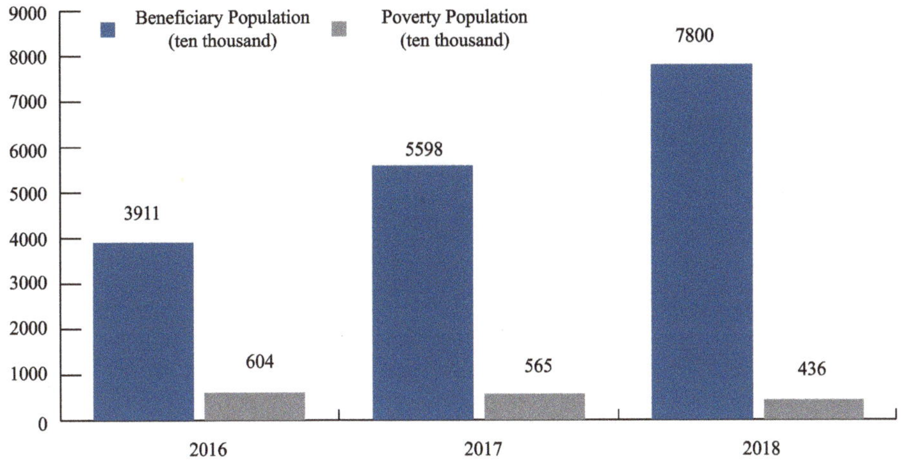

water sources has improved the living conditions in rural areas, improved the quality of life of rural residents and promoted the construction of a beautiful countryside. The central fund subsidy for rural drinking-water safety projects mainly gives priority to minority areas, border areas, pastoral areas and other economically underdeveloped areas. Preferential supporting policies are given to temples, monks and other religious places. Through the implementation of rural water supply projects, tap water is supplied, which not only promotes the coordinated development of ethnic areas and other regions and the equalization of basic public services in urban and rural areas, but also promotes the unity and integration of different ethnic groups and maintains stability in border areas.

Since 2018, the satisfaction rate of water users has exceeded 94% in many investigations conducted by the Ministry of Water Resources. The rural water supply project has become one of the most popular projects benefiting the rural residents.

2.2 Key Experiences and Practices

2.2.1.Scientific Planning, Government-led Promotion

Since 2000, the central and local governments at all levels have carried out top-level design, scientific planning and high-level promotion. With the approval of the State Council, the Ministry of Water Resources implemented the National Tenth Five-Year Plan for Solving Rural Drinking Water Problems in 2000. By the end of 2004, a total of 60.04 million rural residents had their serious drinking water problems solved.

Ceremony for Well Completion and Water Discharge in Sunnan County, Gansu Province

Since 2005, the State Council has approved the implementation of the *Emergency Plan for Rural Drinking-water Safety During* 2005-2006, *Eleventh Five-Year Plan for the National Rural Drinking-water Safety Project* and *Twelfth Five-Year Plan for the National Rural Drinking-water Safety Project*. By the end of 2015, the problem of drinking water safety for 520 million people in rural areas and more than 47 million teachers and students has been solved. These programs, implemented on the basis of existing water supply network, have basically solved the permanent problem of water safety in rural areas.

In order to further improve the level of rural water supply safeguarding, starting from 2016, with the consent of the State Council and under the guidance of the Ministry of Water Resources and relevant ministries and commissions, all provinces (autonomous regions and municipalities directly under the Central Government) have compiled and implement the Plan for Rural Drinking-water safety Consolidation and Improvement during the 13th Five-Year Plan Period. It was approved by the provincial people's governments, with the focus on solving the drinking water safety problems of poverty villages and people. It insists on the large-scale construction of water supply projects, strengthening water-source protection, project management and protection, and comprehensively using transformation, matching, upgrading, networking and other ways to consolidate and improve the level of rural water supply safeguarding.

2.2.2 Differentiated Investment Policies due to Regional Differences

In order to ensure that the rural population of all ethnic groups in all regions of China share the fruits of reform and opening up, central funds are tilted to ethnic minorities, border areas and economically underdeveloped central and western regions. During 2005-2015, 45% of central funds were used to support the western region (the population in the western region accounted for 28%), and 85% of central funds were used to support the central and western regions. On the basis of the implementation of the *12th Five-Year Plan of the National Rural Drinking-water Safety Project*, in order to further solve the drinking water problem in Tibetan areas, pastoral areas and areas with special difficulties, the Chinese government invested 3.621 billion yuan in Tibetan areas, plateau areas, Xinjiang and surrounding areas in 2015. It solved the drinking water problem of 5.666 million people of the rural population in areas mainly inhabited by ethnic minorities. From 2016 to 2020, central funds will continue to increase subsidies, 57.9% of which will be used to subsidize projects to consolidate and improve rural drinking-water safety in western provinces.

2.2.3 Focusing on Targeted Poverty Alleviation and Giving Priority to the Construction of Drinking Water Projects for Poor People

During the 13th Five-Year Plan period (2016-2020), in accordance with the central government's goal of ending poverty and building an equally well-off society, the Ministry of Water Resources has regarded rural drinking-water safety as the number one project for poverty alleviation and has made full efforts to promote it. It has cooperated with relevant departments to increase the support from central funds in areas of poverty, especially those in severe areas of poverty, and has allocated 24.515 billion yuan in subsidies to the provinces that have the task of getting out of poverty through rural drinking-water safety. Provincial governments have increased their investment and have generally included rural drinking water safety protection and poverty alleviation in the scope of assessment for city and county Party Committees, governments or water conservancy departments. In some areas, people's governments at all levels have signed letters of responsibility to further confirm their commitment to push forward the construction of consolidation and upgrades of rural drinking water safety project. In 2018, the Ministry of Water Resources organized the compilation of *Guidelines for of Rural Drinking Water Safety Assessment* to define the assessment criteria for poverty alleviation in rural drinking water safety, according to local conditions in various provinces. In conjunction with the Poverty Alleviation Office of the State Council and the National Health Commission, the Ministry of Water Resources issued the *Notice on Resolutely Winning the Battle for Poverty Alleviation through Safe Drinking Water in Rural Areas* to deploy the work. The Ministry of Water Resources also requested all localities to incorporate the construction of rural drinking-water safety projects for poor villages and people into the county poverty alleviation project. Localities were also requested to prioritize funding and implementation. Measures are in place to ensure that there is not one village or one person among the poor affected by unsafe drinking water. The Ministry of Water Resources has, in conjunction with water conservancy departments around the country, established working ledgers for poor people with drinking water problems from county to village to household. The Ministry of Water Resources has also implemented dynamic identification and rolled out numbers on a monthly basis, to ensure a comprehensive solution to the drinking water safety problems of poor

people by the end of 2020.

2.2.4 Strengthening Management and Protection, Promoting Long-term Operation of the Project

The Ministry of Water Resources, in conjunction with relevant departments, has instructed all localities to continuously improve and optimize the long-term operation mechanism of the project while promoting the construction of rural water supply projects. One is to establish a responsibility system of management. At the beginning of 2019, the Ministry of Water Resources issued the *Notice on Establishing a Responsibility System for Rural Drinking water safety Management*, deploying all localities to fully implement the "Three Responsibilities": the independent duty of local people's government, the industry supervision responsibility of the water administration departments and the operation and management responsibility of water supply units. By June 2019, the "Three Responsibilities" had been established. The second is to establish a central financial subsidy mechanism. In 2019, the Ministry of Water Resources and the Ministry of Finance have arranged 1.45 billion yuan of central subsidy funds to support the maintenance of rural water supply projects in areas of poverty in the central and western regions. They have jointly issued guidance to standardize the use and management of funds, promoted the establishment of incentive mechanisms and speeded up the normal operation of projects. The third is to strengthen industry supervision. Since 2018, the Ministry of Water Resources has organized the agencies of river basin management and other units to conduct large-scale unannounced visits and inspections in accordance with the general principle of "Accelerating the construction of water conservancy projects and strengthen the supervision of water conservancy industy". Provinces have also increased unannounced visits and inspections and established a regular unannounced visit mechanism. In order to maximize the supervisory role of the masses, the telephone numbers of the Ministry of Water Resources, provincial, municipal and county-level agencies for reporting and supervising rural drinking water have been opened. The fourth is to establish a list of problems

and solutions. For rural water supply problems discovered through various channels such as media exposure, unannounced visits and checks, reports from the masses, etc., a list of problems and work ledgers shall be established, and "one province, one list of responsibilities" shall be issued in a timely manner. All localities shall be required to rectify problems one by one. If necessary, they shall "look back" and keep a close eye on implementation, so as to ensure that each piece is implemented and everything has an echo.

2.3 Challenges

Due to China's national conditions, water regime, regional differences and the residential characteristics of the rural population, in many areas there is still a large gap between the level of rural water supply and the needs of the masses for a better life. With socio-economic development, the demand for water has increased in all aspects, and the amount of water coming from drinking water sources in many rivers, lakes, reservoirs and villages has obviously decreased. Especially in dry seasons and water use peak periods, the guarantee rate of water has continuously decreased. Northwest China is dry and short of water and there is no major water sources. Southwest China has insufficient water storage and supply capacity. There are many rural water supply projects with a wide scope. Some of them are small in scale, with scattered and unstable water sources. Some of them have not established a reasonable water price and water fee collection mechanism. In particular, there is no high rate of water fee collection for some small-scale projects, and in others none at all. Maintenance funds are generally insufficient. There is a lack of professional management talent and teams, so the operation and management of the projects is difficult.

3. Future Goals

According to the requirements of "resolutely winning the battle against poverty and building a well-off society in an all-round way", implementing the strategy of rural revitalization and the development of urban-rural integration, we fully respect the wishes of farmers, adjust measures to local conditions and adhere to the principle of balancing urban and rural areas. We also put equal emphasis on construction and management, establish and perfect the rural water supply project and management system from the source to the water tap, and further improve the penetration rate of rural tap water, water quality compliance rate and management of project operation.

By 2020, the ratio of access to the centralized water supply for rural residents should be over 87% and the coverage ratio of tap water in rural area should be over 83%; water quality will rise dramatically in rural areas. With the improvement of operation, management and maintenance system for rural water supply projects, water supply in rural areas will step into sustainable development. By 2025, it is expected that the coverage ratio of tap water in rural area should be 88% and the water quality compliance rate will be significantly improved. The proportion of the population served by the extension of urban water supply network and large-scale water supply projects with more than 10,000 people will be 55%.

4. International Cooperation and Exchanges.

Before large-scale construction of safe rural drinking-water projects conducted

Drinking Water from the Newly-drilled Well

by the Chinese government in 2005, China has received help and support from such international organizations as the United Nations Children's Fund (UNICEF), the United Nations Development Program (UNDP), the World Health Organization (WHO), the World Bank and the Asian Development Bank (ADB) as well as the governments of countries such as the United Kingdom to ensure access to safe drinking water for rural residents. In the 1980s, China implemented multiple projects of rural water supply and environmental sanitation with credit and lending from the World Bank. In the 1990s, China and UNICEF explored and applied in some areas with success a "3 in 1" approach that features low cost, sustainable water supply, environmental sanitation and health education. Subsequently, China and the UK have worked together to develop a method that allows water users to fully participate in rural water supply projects. The method is currently being disseminated in rural areas of China.

In recent years, China has undertaken a number of water safety projects in the developing world by sending technical experts to numerous African nations where the Chinese specialists guide their host countries in their efforts to build water supply projects. In addition, China has run training programs for a large number of technicians and managers from these countries.

A mother in Remote Mountain Areas Gets Access to Tap Water

Chapter 7
Irrigation and Drainage

Written by: Dang Ping
Translated by: Shi Xianglin

1. Overview

China is a country with the largest area of irrigated land, and much of it is used for agricultural purpose. Farmland irrigation and drainage in China has a history of more than 5,000 years, from Great Yu Controlling the Waters until now. It plays a decisive role in Chinese agricultural production, food security as well as economic and social development. A large number of ancient irrigation and drainage projects such as the world-famous Zhengguo Canal, Dujiang Weir and Lingqu Canal are effective to this day, providing economic, social and ecological benefits.

Karez Well in Xinjiang

Since the 1950s, the Chinese government has focused on the requirements of economic and social development. It has organized people of all ethnic groups across the country to carry out large-scale water conservancy construction. This has helped with flood control and drought relief, guaranteeing

A Wheat Harvest in the Irrigated Area of the Lower Reach of the Yellow River

water supply and food security, and helped with ecological improvement. The irrigated area has been continuously expanded increasing its support capacity. As the basic conditions for agricultural production constantly improved, new breakthroughs in grain output continue to be made.

2. Main Achievements & Challenges

2.1 Main Achievements

2.1.1 Farmland Irrigation and Drainage System is Gradually Improved

Through the construction and reinforcement of reservoirs and other water storage projects and the implementation of inter-basin and cross-regional water diversion projects, the water resources safeguard system has been initially formed. The system combines storage, diversion and transfer and effectively alleviates the problem of an uneven distribution of water resources. Thus farmland irrigation and drainage facilities are upgraded and the capacity for agricultural production

Headworks of Dujiangyan

is greatly improved. At present, 7,800 reservoirs in various types have been built, and each has an designed irrigation capacity of 667 hm^2 ; 4.6 million small dams and 6.9 million water cellars have been completed, which are supplemented by 4.9 million electromechanical farmland wells with an installed capacity of 49.5 million kW; 434 thousand fixed irrigation drainage pumping stations have been constructed, with an installed capacity of 27.16 million kW; The size of irrigated arable land has grown from 16 million hm^2 in 1949 to 68 million hm^2 at present, ranking first in the world. The total grain output has gone up from 113.2 billion kg in 1949 to 650 billion kg. More than 75% of grain is produced in irrigated areas.

2.1.2 Water-saving Irrigation is a Key Water Conservancy Task for Sustainable Utilization of Water Resources

Water for agricultural use accounts for about 60% of the national economic and social water consumption. The Chinese government put forward that "water-saving irrigation should be carried out as a revolutionary measure". The local governments at all levels and all sectors of society, especially the farmers who had benefited from this measure, have actively participated in the construction and management of water-saving irrigation projects. Through the continued construction of irrigation facilities and implementation of water-saving measures in irrigated areas, China's water-saving irrigation has been developed. The coverage of water-saving irrigation projects has increased to 35.33 million hm^2, including 22.27 million hm^2 of high-efficient water-saving irrigation through low-pressure pipe irrigation, sprinkler irrigation, micro-irrigation and so on. The effective utilization coefficient of irrigation water in farmland, which is regarded as an indicator of efficient water use in irrigation, increased from 0.43 in 2000 to 0.554. In the past 30 years, the area of effectively irrigated farmland in China has increased by 20 million hm^2 .Water consumption for agricultural irrigation has basically remained at about 350 billion m^3, but the proportion of water used for irrigation has decreased from more than 79% to the current 55% or so. At present, China's water-saving irrigation measures have developed from irrigation in small areas to concentrated contiguous irrigation in large areas, from field water saving to comprehensive water saving, from plain well irrigation to gravity irrigation and arid hills irrigation, and from efficient cash crops irrigation to field crop irrigation.

Sprinkler irrigation

2.1.3 The Supporting System for Irrigation and Drainage is Gradually Improved

The *Water Law of the People's Republic of China, Regulations for Irrigation and Drainage* and *Opinions on Promoting a Comprehensive Reform of Water Price for Agriculture* requires the government to take irrigation and drainage construction seriously. The Chinese government has formulated the *Rural Revitalization Strategic Plan* and the *National Economic and Social Development Plan*, which clearly define the objectives and tasks of the construction and reconstruction of farmland irrigation and drainage facilities and put forward developmental requirements, which are the focus of the national infrastructure. We have established an investment and financing system for irrigation and water conservancy where public finance is the leading factor, with social supplements and farmers' participation. We have also improved supporting policies, either civilian-run or state-subsidized, such as replacing subsidies with awards, providing subsidies for the purchase of agricultural machinery and tools, and preferential financial loans. The Chinese government has issued and implemented the stringent water resources management system, the *National Water Saving Action Plan*. It has formulated a system for farmland water conservancy technology, product and equipment standards and various other project construction management standards. As China's scientific and technological capacity has significantly improved, a water-saving technology system suitable for the national conditions has emerged. The aim to integrate an industrial system for equipment and materials production with a water-saving system combined with scientific research and application is well on the way.

2.2 Challenges

2.2.1 Ensure Food Security and Comprehensively Build a Well-off Society

With a huge population of nearly 1.4 billion, it is always a top priority for China to solve the problem of feeding the population while relying only on domestic resources. Therefore, we must speed up the construction of farmland water conservancy facilities and strive to improve the total agricultural production capacity and output efficiency. Then we can realize the aim of "storing grain in land and technology".

Pishihang Irrigation District, Located in Anhui Province

2.2.2 Relieve the Contradiction between Supply and Demand of Water Resources and Promote Ecological Civilization

At present, the average annual water shortage of China's agriculture is about 30 billion m^3. In some agricultural areas, the following problems are obvious: extensive management, inefficient water consumption, over-exploitation of groundwater and transfer of water originally supplied for ecological purpose. We must vigorously develop water-saving irrigation, adjust the agricultural layout and planting structure, change the traditional way of agricultural development, promote water-saving technologies and scientific irrigation methods, and enhance the efficiency and economic returns of agricultural water consumption.

2.2.3 Adapt to Rural Reform and the Development of Agricultural Modernization

At present, China's rural economic and social structure is undergoing profound changes. The rural labor force has shifted to the cities and is ageing. The "Empty-nest" phenomenon is becoming more prevalent. The benefit-cost ratio of agriculture is comparatively low. We must vigorously accelerate the modernization of irrigation and drainage, through information technology and scientific endeavour, with the aim to effectively reduce labor intensity while improving productivity.

2.2.4 Cope with Global Climate Change and Improve the Capacity to Reduce the Losses Resulted from Drought and Other Natural Disasters

Affected by climate change, the impact of extreme flood and drought on agricultural production is increasing. It is necessary to reinforce the reconstruction of irrigation and drainage facilities, improve the reliability of agricultural water supply and drainage, then lay a solid foundation for water conservancy to enhance the agricultural sector's ability to effectively cope with climate change and to combat the effects of floods and droughts.

3. Strategies and Key Measures for Future Development

3.1 Developmental Goals of Irrigation and Drainage in China by 2030

3.1.1 Extend irrigated area to 76.3 million hm^2 nationwide, which will include 71.3 million hm^2 of effectively irrigated farmland; and basically restore and rehabilitate farmlands and arable land throughout the country that is vulnerable to being waterlogged or salinization.

3.1.2 Extend the coverage of the national farmland water-saving irrigation project to more than 56.70 million hm^2, which will include 34 million hm^2 by sprinkler irrigation, micro-irrigation, or pipeline irrigation.

3.1.3 Increase the farmland irrigation water use coefficient to 0.6; and cap nationwide water consumption for irrigation purposes at 373 billion m^3 .

3.1.4 Tighten irrigation water use management; establish a system that features "total quantity control and quota management" of irrigation water consumption; and set up a sound incentive and constraint mechanism.

3.2 Priority Tasks

3.2.1 Speed Up the Transformation and Construction of Major Irrigation and Drainage Engineering Facilities

Complete the construction of supporting facilities and water-saving renovation in

Dujiangyan Irrigation District

large and medium-sized irrigation districts nationwide. Speed up the renovation of irrigation and drainage pumping stations, strengthen the management of key waterlogged areas, and improve the irrigation and drainage system. We will comprehensively promote the modernization and reconstruction of large and medium-sized irrigation districts and build a number of new large and medium-sized irrigation districts in areas where water and soil resources permit.

3.2.2 Develop Water-saving Irrigation Vigorously

Speed up the implementation of efficient water-saving irrigation on a gigantic scale in the main grain producing areas, water shortage areas and ecologically vulnerable areas; vigorously promote pipeline irrigation, sprinkler irrigation, micro-irrigation and other efficient water-saving irrigation technology; enhance water-saving in agriculture through a combination of engineering measures, advanced agronomic techniques and improved management.

3.2.3 Comprehensively Push Forward the Construction of Small Irrigation and Drainage Projects

Construct field work in irrigation areas and small-scale projects in mountainous areas, such as the "Five Kinds of Small Water Conservancy Projects", including small dams, small pumping stations, small channels, small pools and small water cellars; and speed up the construction of drought-resistant emergency water projects in the flood-prone areas so as to improve the support capacity of anti-drought projects.

3.2.4 Quicken Reform in the Field of Irrigation and Drainage

Continue to reform the management of pumping stations and other key irrigation and drainage projects in irrigation areas; reform the property rights system of small-scale farmland irrigation infrastructures; facilitate the development of the "water users association" among farmers; and actively cultivate specialized and marketed service teams. By 2020, the irrigation area where farmers participate in the management of irrigation water will reach more than half of the effective irrigation area. Actively carry out the total control and quota management of irrigation water; construct water metering and informatization facilities; and reform agricultural water price.

Small Waterworks in Guizhou Province

4. International Cooperation and Exchanges

Since the 1980s, we have engaged in both multilateral technical exchanges and cooperation with international organizations including UNFAO, UNICEF, the EU, the World Bank, Asia Development Bank, the International Water Management Institute, and the International Commission on Irrigation and Drainage, as well as bilateral exchanges and cooperation with countries such

Nigerian Engineers Visited Water Tank in Dingxi City, Gansu Province

as Japan, the UK, Australia, the Netherlands, and so on. For instance, we cooperated with the Japanese government in execution of the "Development and Training Center for Irrigation and Drainage Technology" program and the "Demonstration Project on Water-saving Renovation of Large Irrigation Areas in China". In cooperation with the UK, we carried out the World Bank managed project "Reforming Rural Water Conservancy in China for the Poor". Economic or technical cooperation between China and other developing countries in Asia and Africa also goes on in areas such as the construction of irrigation and drainage projects, rural water supply facilities and personnel training. This plays a positive role in helping those countries improve their rural and agricultural infrastructure, increase grain output, and raise living standards of the local population.

As a member of the International Commission on Irrigation and Drainage (ICID), China has conducted all-round exchanges with the ICID, organized several international irrigation and drainage conferences, participated in the work related to the ICID, and carried out bilateral cooperation and exchanges with ICID member countries. One Chinese expert was elected as ICID Chairman, five were elected as vice chairmen, and 13 were granted with ICID water-saving technology awards. Water-saving Nigerian youth-engineers

ICID Heritage Irrigation Structures in China

Sprinkler Irrigation in Guangxi Zhuang Autonomous Region

visited the water tank in Dingxi City, Gansu Province. China also won the ICID Heritage Irrigation Structures in China Headworks of Yintang Irrigation District in Heilongjiang Province award and water management award. As of 2019, China has 19 ancient irrigation projects listed in the ICID Register of Heritage Irrigation Structures.

Chapter 8
Soil and Water Conservation

Written by: Zhang Chao
Translated by: Shi Xianglin

1. An Overview of Soil and Water Erosion

China is one of the countries with the most severe soil and water erosion in the world. In 2018, the areas of soil and water erosion around the country reached 2.7369 million km^2, accounting for 28.6% of the national territorial area (excluding Hong Kong, Macao and Taiwan), including 1.1509 million km^2 of areas with water erosion and 1.5860 million km^2 of areas with wind erosion. There is no doubt that severe soil and water erosion results in land and water resource destruction, ecological environment deterioration, intensification of natural hazards. It also creates threats to national ecological safety, flood control safety, drinking water safety and food safety, so it is a prominent constraint on the sustainable development of China's economy.

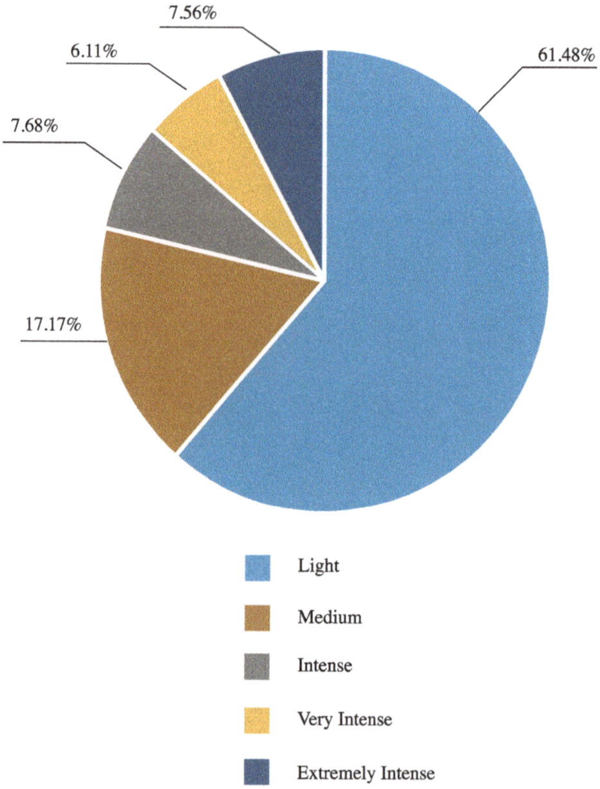

Percentage of Soil and Water Erosion at Each Intensity around the Country

2. Major Achievements

China always attaches importance to soil and water conservation and has carried out large-scale soil and water erosion control, achieving remarkable results. According to data from the Ministry of Water Resources, the area of soil and water erosion around the country in 2018 was reduced by 933,400 km^2 in comparison with 1985, nearly a decrease of 10%. Compared with the first national water resources census in 2011, the area was reduced by 212,300 km^2 with a decrease of 8.2%. The erosion areas above medium level decreased by 511,100 km^2 with a decrease of 32.65% in comparison with 2011. At present, the area ratio of medium and light soil and water erosion is 78.7%. Compared with 2011, the area of soil and water erosion in the Three Gorges Reservoir Region, black soil region of northeast China, Danjiangkou Reservoir Area and its upper reaches, Beijing-Tianjin-Hebei Region, Loess Plateau, southwestern rocky desertification area, and the Yangtze River Economic Zone respectively decreased by 18.63%, 12.42%, 11.85%, 11.56%, 9.13%, 8.88% and 8.86%. Their rate of decline was higher than the national level. As a whole, the area and

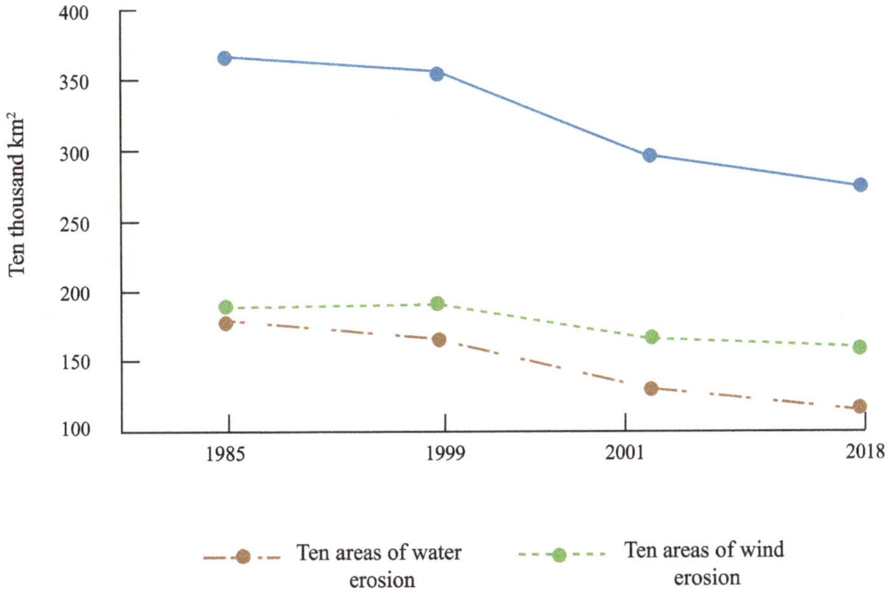

The National Area Changes of Soil and Water Erosion (1985-2018)

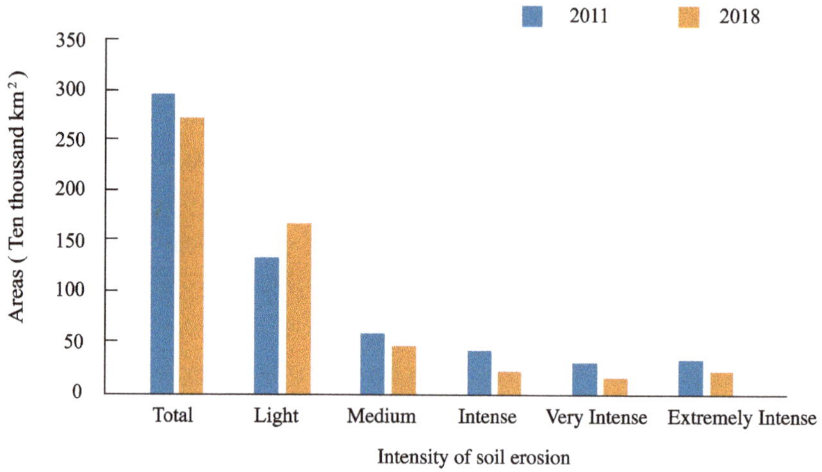

The National Area Variation of Soil and Water Erosion

severity of soil and water erosion in China is reducing year by year. This is a remarkable improvement.

2.1 Comprehensive Implementation of *Soil and Water Conservation Law* and Effective Control of Artificial Soil and Water Erosion

In 1991, the *Soil and Water Conservation Law of the People's Republic of China* was officially implemented, so that soil and water conservation work in China is gradually on the way to prevention and control according to the law. Executing a "three-simultaneous" system (soil and water conservation facilities in construction projects must be simultaneously designed, constructed and put into use with main construction) of soil and water conservation in production construction projects has achieved significant results. In 2011, the revised *Soil and Water Conservation Law* came into effect, adding provisions to the main responsibility of the government to legally strengthen prevention and protection policies and accountability. It provided a more powerful legal safeguard for protecting water resources according to law. The administrative department in charge of water resources at each level seriously fulfils its responsibilities of supervising soil and water conservation in production and construction projects. These include supervising artificial soil and water erosion caused by production and construction activities, examining and approving soil and water conservation schemes, and source management. Production and construction projects that do not meet the requirements of ecological protection and water and soil conservation will not be approved. We will earnestly strengthen the follow-up inspection of the implementation of the water and soil conservation plan, establish and improve a step-by-step supervision system for water administrative departments to perform their duties according to the law, strengthen supervision during and after the event, and strictly investigate and punish illegal acts. At the same time, we will make full use of remote satellite imaging, unmanned aerial vehicles and other information-based means to innovate supervision methods and improve supervision efficiency. Since the promulgation and implementation of the *Soil and Water Conservation Law*, more than 540,000 production and construction projects nationwide have implemented soil and water conservation programs and measures, and reduced the area of human-induced soil and water loss due to development and construction by 220,000 km^2 .

2.2 Continuous Implementation of Comprehensive Soil and Water Erosion Governance and Remarkable Improvement of Production and Life Conditions and Ecological Environment in Governed Areas

Batches of key projects of soil and water conservation have been successively launched and implemented in the midstream of the Yellow River, the upper Yangtze River, the silt dam of the Loss Plateau, sources of sandstorms in Beijing and Tianjin, the black soil region of Northeast China, and rocky desertification control in Karst regions. The range of control has expanded to major river basins nationwide from the traditional upper and middle reaches of the Yellow River and the Yangtze River, basically covering areas with serious soil erosion. We have effectively promoted a series of major projects such as the three northern (northwest-north-northeast) shelter-belts, the conversion of farmland to forests and grassland, the protection of natural forest resources, the rehabilitation of land, desertification and rocky desertification, and the ecological protection and restoration of mountains, rivers, forests, fields, lakes and grassland areas. At the same time, due to improved policy and operation as well as the encouragement of social forces to participate in soil erosion control, the whole society jointly controls soil erosion. By the end of 2018, the country has initially controlled an

Combating desertification in the Shule River basin, Gansu Province

area of 1.31 million km^2 of soil erosion. The existing soil and water conservation measures can reduce the amount of soil erosion by 1.6 billion tons per year. The production and living conditions in the controlled areas have improved significantly, the grain output has increased, the income of farmers has increased significantly, and the countryside has taken on a new look. In the 21st century, the Ministry of Water Resources has further expanded its scope of work and gradually promoted the construction of ecologically-clean small watersheds around the new rural construction and rural revitalization strategy. This has produced valuable experience in improving the rural landscape and preventing non-point source pollution.

2.3 Attention on Developing Self-repairing Capacity of Ecology and Rapid Recovery of Vegetation in Large Areas

After entering the 21th century, China's water and soil conservation methodology has been undergoing constant reform. While strengthening the management of water and soil loss, it has implemented enclosure protection, banned grazing and rotational grazing. It has fully relied on nature's self-repairing ability to speed up vegetation restoration, reduce water and soil loss and improve the ecological environment. The Ministry of Water Resources has successively launched two batches of pilot projects for the ecological restoration of soil and water conservation in more than 200 counties in 29 provinces (autonomous regions). Furthermore, it has implemented prevention and protection projects

Soil and water conservation along a highway in Baise City, Guangxi Zhuang Autonomous Region

for soil and water conservation in the "Three Rivers Source" area of Qinghai Province. 136 cities and nearly 1200 counties in 27 provinces, autonomous regions and municipalities directly under the central government have implemented mountain closures and grazing bans. Key national water and soil conservation project areas have implemented full-scale enclosure protection. Many practices have been spread around the country, including grass for livestock, construction, modification and relocation for repair, and energy substitution. This method of ecological recovery has gradually matured, attaining a good ecological effect. The philosophy of sticking to prevention first, followed by crucial protection and the full development of nature's capacity to self-repair enjoys popular support.

2.4 Acceleration of Supervision and Information Application and Improvement on Soil and Water Conservation Management Efficiency

The Ministry of Water Resources has carried out three national remote sensing surveys of soil and water loss, basically finding out the situation and dynamic trend of soil and water loss around the country. The second phase of the construction of the national soil and water conservation monitoring network and information system has been implemented. 175 monitoring substations and 738 monitoring points have been built, initially forming a monitoring network system covering the major types of soil and water loss areas in the country. The monitoring and forecasting capabilities of soil and water loss have been significantly enhanced. Since 2003, it has consecutively issued soil and water conservation bulletins for 15 years, causing an important impact on society. Since 2008, regular dynamic monitoring of soil erosion has been carried out in key prevention areas and control areas for soil erosion. In 2018, the annual full coverage of 9.6 million km^2 of land area has been achieved for the first time, and the status of soil erosion from the whole country to county-level administrative areas has been quantified. The development of the national water and soil conservation information management system has initially established a platform for information development. This includes the comprehensive application of information technology in supervision and management. Comprehensive management, monitoring and evaluation has been continuously increased, effectively improving the management level and efficiency of soil and water conservation.

2.5 Promotion of Plans, Standards and Technical Innovation and Remarkable Enhancement of Basic Supporting Capacity in Soil and Water Conservation

In 2015, the State Council approved the *National Soil and Water Conservation Plan (2015–2030)*. 31 provincial soil and water conservation plans have been approved by the provincial-level people's government or the authorized department. Since then, special plans such as the comprehensive treatment of erosion gullies in the black soil region of Northeast China and comprehensive treatment of soil erosion on sloping farmland have been made and approved one after another. The top-level design of soil and water conservation has become increasingly perfect. What's more, a number of national soil and water conservation scientific research stations, test areas and key laboratories on soil erosion have been built, and significant progress has been made. A large number of major scientific and technological research projects on soil and water conservation have been undertaken. In conjunction with the Chinese Academy of Sciences and the Chinese Academy of Engineering, a comprehensive scientific investigation on soil and water loss and ecological safety in China has been carried out. 130 national soil and water conservation science and technology demonstration parks have been built. Practical soil and water conservation technologies focusing on slope-to-terrace, slope river system and rainwater utilization have been tested and promoted. The contributions of science and technology to soil and water conservation has continuously increased. A water and soil conservation standards system covering the three categories of integration, construction and management and 14 additional functions has been established. There are currently 60 effective water and soil conservation standards. A technical standard system that meets China's national conditions and the needs of water and soil conservation work has been established, providing basic support for the prevention, control and supervision of water and soil loss.

3. Overall Working Ideas and Important Measures

3.1 Overall Ideas

It is necessary to adhere to the problem-oriented and goal-oriented approaches, pay close attention to the implementation of responsibilities, focus on strengthening the supervision of man-made soil erosion, focus on improving policy implementation, focus on strict supervision and accountability, fully rely on advanced technical means, perform the duties of soil and water conservation, strive to improve management capacity and level, provide strong supervision and effective governance, and provide support for accelerating the construction of an ecological civilization by ensuring sustainable economic and social development.

Silt Retention Dam Built in Jiuyuangou River Basin in Shaanxi Province

By 2035, soil and water loss in key prevention and control areas will be comprehensively controlled. Man-made soil and water loss will also be managed. The area and intensity of soil and water loss will be greatly reduced. The quality and benefit of soil and water loss control will be significantly improved. The national soil and water loss situation will be fundamentally improved, and the modernization of soil and water conservation control systems and capabilities will be realized.

By 2050, the soil and water conservation

management system and management capacity will be modernized, and the soil erosion situation around the country will be comprehensively and effectively controlled, providing strong support for building a beautiful China, an ecological civilization and realizing common prosperity for all people.

3.2 Important Measures

3.2.1 Implementation of the Toughest Soil and Water Conservation Control

Supervision and management of soil and water conservation is an important responsibility entrusted to water administration departments at all levels by the *Soil and Water Conservation Law*. At present and for a period to come, we should set up a systematic, clear-cut, strict, efficient, standardized and orderly supervision system with the goal of monitoring man-made soil and water loss. We should implement the strictest prevention of soil and water loss, the strictest examination and approval of soil and water conservation plans, the strictest in-and-out supervision of soil and water conservation, and the strictest law enforcement of soil and water conservation. We should enact comprehensive administrative punishment, credit punishment and other means to strictly investigate and deal with violations of laws and regulations related to soil and water conservation.

3.2.2 Acceleration of Developing Soil and Water Erosion Control

Considering mountains, rivers, forests, farmlands, lakes and grasslands as a whole, it is necessary to accelerate and facilitate comprehensive control of soil and water erosion, maximize the comprehensive benefits of soil and water conservation, and to meet the growing needs of people for a better life. Focusing on the black soil areas in the upper and middle reaches of the Yangtze River, Yellow River and in the Northeast, we will accelerate the improvement of sloping farmland, erosion gullies and small watersheds. We will actively promote the construction of ecologically clean watersheds according to local conditions, and build a soil erosion control system that meets the requirements of construction for an ecological civilization in the new era.

3.2.3 Construction of the Scientific and Complete Soil and Water Conservation Monitoring Service System

Monitoring is considered as the foundation of soil and water conservation. In the new era we should focus on the needs of reinforcing supervision and improving weaknesses, and fulfilling the government's responsibility for soil and water conservation monitoring by constantly optimizing the dynamic monitoring of annual soil and water erosion. To aid with this we should construct a monitoring network with reasonable layout, complete functions and coordination between upper and lower levels. We must also timely and accurately master the dynamic changes of soil and water erosion in county-level administrative districts and key areas around the country, positively push the promotion and application of high-tech products and processes in the monitoring field, reinforce the application of monitoring results, realize the effective connection between monitoring and management, and offer powerful support to the assessment of soil and water conservation targets. These are the red line warnings of ecological safety for soil and water conservation.

3.2.4 Great Promotion of Information Technology in Soil and Water Conservation

Informationization is the leading power to guide innovation and drive transformation. The task of soil and water conservation in the new era should be to promote the applications of information technology in soil and water conservation. This would be establishing the national soil and water conservation information system and database with mutual connection and shared resources, intensifying internal and external data shared in the industry, and accelerating

Terraced Fields in Yuanyang County, Yunnan Province

the full application and in-depth integration between high-tech information technology, supervision, governance and monitoring. This will allow us to supervise production and construction activities, control new soil and water erosion, reveal any effects of new soil and water erosion in a timely manner, and comprehensively improve the modernization level of soil and water conservation.

4. International Cooperation and Exchanges.

China hosts the office of the Secretariat of World Association of Soil and Water Conservation (WASWAC, http://www.waswac.org/) which currently consists of 1,335 private members and 10 unit members in 85 countries and regions. In 2010, Chinese experts were elected President of the WASWAC Council and Secretary General of WASWAC. In April 2015, the Chinese government officially approved the registration of the WASWAC. The WASWAC organizes a series of international symposiums every three years, with a quarterly publication of the Journal of *International Soil and Water Conservation Research*, and a monthly publication - WASWAC Hot News. China also holds the presidency of the International Seabuckthorn Association(ISA, http://www.isahome.net/) and is home to the office of the ISA's Secretariat. The ISA represents a key player in the field of international seabuckthorn ecological improvement and industry development. China has successively played host to a number of important conferences including the 12th International Soil Conservation Organization Conference (ISCO 2002), the 2nd and 5th International Seabuckthorn Association Conference, the International Symposium on Prevention and Control of Soil Degradation that concurrently served as the 1st ISCO Annual Conference, the International Youth Forum on Soil and Water Conservation, the China-US Workshop on Soil and Water conservation, and the China-Africa Workshop on Soil and Water Conservation. China has strengthened multilateral cooperation and exchanges by pursuing

Chapter 9
Flood and Drought Disaster Prevention

Written by: Li Tieguang

Translated by: Xiao Rui and Shi Xianglin

1. Overview

1.1 Main Types of Flood Disasters in China.

1.1.1 Basin-wide Floods

The regions seriously threatened by floods in China are around the Yangtze River, Yellow River, Huaihe River, Haihe River, Pearl River, Songhuajiang River, Liaohe River, rivers flowing into the sea alone and Taihu Lake, where there is 1/2 of the national population, 1/3 of the national arable land and 3/4 of the gross industrial and agricultural output of China. Since 1949, more than 50 major basin-wide floods have hit China. It is expected that, by 2020, a total population in the areas protected by flood control measures in the middle and lower reaches of the major rivers will be 600 million and the GDP will be RMB 26 trillion.

Mouth of Mudslide Valley in Zhouqu Disaster-hit Area in Gansu Province

1.1.2 Flash-floods

The area in China directly threatened by flash-floods extends to 480,000 km^2 with a population of more than 76 million people. The area in need of flash-flood prevention and control in China is 3.86 million km^2 with a population of 300 million people. The death toll of flash-floods usually accounts for 70% of the total casualties caused by flood disasters.

1.1.3 Urban Flooding

637 cities in China are tasked with flood control. Almost all of them have preparatory plans for flood control. 490 cities have finalized their urban flood control plans, 355 cities completed initial design of flood control works, and 321 cities reached the national standards for flood control. The completed 28,000 km of urban dikes and embankments protects an urban area of 88,000 km^2. In recent years, water-logging in urban areas has worsened.184, 243, 125, 164 and 192 cities (including county towns) in China were severely logged in 2012, 2013, 2014, 2015 and 2016 respectively, which resulted in immobilized transportation in certain urban districts, severe impacts on the normal operation of the cities, and enormous economic losses.

1.2 Drought

1.2.1 Agricultural Drought

Since 1949, the average drought-affected area per annum in China is 20.66 million hm^2, whereas average annual grain loss stands at 16.35 million metric tons or even goes up to 60 million metric tons in extremely dry years. In 1997, when a catastrophic drought hit the Yellow River Basin, the Lijin section in the lower reaches of the Yellow River dried off 13 times for a total of 226 days, and the total length of dried-off river extended over 700 km. In 2000 and 2001, the whole country was caught in consecutive catastrophic droughts. In 2006 Chongqing Municipality and Sichuan Province suffered a 100-year-return drought. In 2009 numerous severe droughts hit the northern part of China. In 2010 another catastrophic drought occurred in southwest China.

1.2.2 Water Shortage in Cities

More than 400 cities in China are short of water, of which over 100 cities suffer severe water shortage. Since 2000, due to consecutive droughts in the Haihe River Basin, Tianjin Municipality was frequently short on water supply. During 2000, the severe drought in Yantai and Weihai in Shandong Province meant available water per capita per month was capped at 1 metric ton in these cities. From the winter of 2004 onward, saltwater intrusion in the estuary of the Pearl River has threatened the safety of water supply in various cities on the Pearl River Delta, including Zhuhai and Macau, etc. Moreover, water pollution has led to frequent urban water supply crises. The drinking water supply for nearly one million inhabitants was suspended in cities, including Neijiang, because of the serious pollution of Tuojiang River in 2004. Jilin and Harbin were both caught in a water supply crisis when the Songhuajiang River was polluted in 2005. Likewise, the cyanobacteria bloom in Taihu Lake in 2007 led to a water supply crisis in Wuxi, Jiangsu Province.

1.3 Prevention and Management Institution of Flood and Drought Disaster and Relevant Laws and Regulations

China has established the State Flood Control and Drought Relief Headquarters, which is responsible for organizing the work of flood control and drought relief throughout the country under the leadership of the State Council. Its general commander is a leader of the State Council who is in charge of the State Flood Control and Drought Relief Headquarters and its members include relevant ministries of the central government as well as the People's Liberation Army (PLA) and the Chinese Armed Police Force (CAPF). The Minister of Water Resources serves as deputy commander-in-chief.

Basin flood control and drought relief headquarters are set up for major river and lake basins (the Yangtze River, Yellow River, Huaihe River, Haihe River, Songhuajiang River, Pearl River and Taihu Lake), and the coordination leading group for flood control and drought relief is established for the Liaohe River basin. It is responsible for flood control and drought relief work within the

Linhuaigang Flood Control Works in Anhui Province

jurisdiction. Local people's governments at or above the county level with flood control and drought relief tasks have set up offices to organize and direct relevant work under the leadership of the higher flood control and drought relief agencies and the people's governments at the corresponding levels. The Ministry of Water Resources is mainly responsible for monitoring and pre-warning of flood and drought, water project scheduling and technical support for flood prevention and fighting.

China has successively promulgated the *Water Law*, the *Flood Control Law*, the *Flood Control Regulations*, the *Drought Control Regulation* and the *Interim Measures for Compensation in Flood Storage and Detention Areas*. All provinces, autonomous regions and municipalities have correspondingly promulgated and implemented the supplementary regulations and rules for flood control and drought relief. In accordance with relevant laws and regulations, various emergency plans for flood control and drought relief, flood prevention schemes for major rivers and their major tributaries and flood dispatching schemes have been formulated. Most of the nearly 100,000 reservoirs and hydropower stations around the country have formulated emergency plans for flood control and scheduling plans for rescue. Preparatory operation plans have been made for all the 98 national flood storage and detention areas, and flash-flood preparatory plans for 2,076 counties with the task of flash-flood prevention.

2. Major Achievements and Challenges

2.1 Major Achievements

Currently, main sections of major rivers in China are by and large capable of withstanding the most catastrophic flood since 1949. With the combined use of dykes, reservoirs as well as flood storage and detention areas, both the Jinjiang section of Yangtze River and the main sections of Huaihe River that are most prone to flood disasters could survive 100-year-return floods; the sections on the middle and lower reaches of the Yellow River are capable of preventing 1000-year-return floods; the rivers in the northern part of the Haihe River basin can defend floods similar to the 1939 one, while the main sections of rivers in the southern part could combat 1963-type floods; and the main sections of Songhuajiang River, Liaohe River, Pearl River and Taihu Lake are basically capable of guarding against the most catastrophic flood since 1949. The dykes along the key coastal areas that used to sustain 5~10-year-return typhoons are now able to defend 50-year-return typhoons. Medium and small rivers are capable of withstanding ordinary floods. When medium drought years appear in some areas, industrial and agricultural production and ecology will not be greatly affected, which can basically ensure the safety of the urban and rural water supply.

China has not only enhanced the construction of a flood and drought prevention system, but also adopted various non-engineering measures across the country. 121,000 hydrologic stations of various types have been built and more than 110,000 flood reporting stations have been set up. The monitoring of hydrologic factors such as water level and rainfall has been fully automated, with the accuracy rate of flood forecasting exceeding 90% in critical periods of major rivers. Flash-flood monitoring and pre-warning platforms have been built in 2,076 counties tasked with prevention and control. Hydrologic monitoring stations and forecasting & pre-warning systems have basically covered all major rivers and 5,218 medium and small rivers for flood control. On the whole, China's major rivers are basically capable of withstanding the most catastrophic flood since 1949. According to the research results of the Chinese Academy of Engineering, cooperation with international organizations including the World Bank, the EU and the Asian Development Bank. The Loess Plateau Watershed Rehabilitation Project of Soil and Water Conservation with the World Bank Loan marks China's first large-scale project using foreign funds to tackle soil and water erosion. The project received the 2003 World Bank President's Award of Excellence. Under the auspices of the Ministry of Water Resources, implementation of the Project of Seabuckthorn Cultivation Demonstration Zone in Bolivia with Chinese Aid started in 2001. Subsequently, the MWR has overseen the implementation of the project of soil and water conservation in the 4 provinces of Yunnan, Guizhou, Hubei and Chongqing with World Bank lending, and the project of small watershed improvement and management with UK grant. Thanks to the rollout of these projects, China's successful experiences in soil and water erosion control and project management have been generalized and found growing acceptance and deployment in other developing nations.

now China's flood control capability has been upgraded to a safer level, and its flood and drought prevention capability has reached an international medium level, which is relatively high in the ranking among developing countries. Thanks to the gradual improvement of engineering and non-engineering systems and the extensive participation of the whole society, flood and drought prevention in China has achieved enormous social benefits and economic returns.

Floods successfully controlled in China in recent years

Year	River/River basin	Scale/Severity of floods
1998	Yangtze River	Basin-wide extraordinary flood
	Songhuajiang River	Basin-wide extraordinary flood
1999	Taihu Lake	Extraordinary flood
2003	Huaihe River	Basin-wide extraordinary flood
	Hanjiang River, Jialingjiang River, Weihe River	Serious autumn flood
2005	Xijiang River (Pearl River basin)	≥100-year-return extraordinary flood
	Liaohe River basin	The most extraordinary flood in recent 20 years
	Hanjiang River, Jialingjiang River, Weihe River	Severe autumn flood
2007	Huaihe River basin	Basin-wide extraordinary flood
2011	Hanjiang River, Jialingjiang River, Weihe River	Severe autumn flood
2013	The lower reach of Heilongjiang River	≥100-year-return extraordinary flood
	Songhuajiang River basin	The most extraordinary basin-wide flood since 1998
	The upper reach of Hunhe River (Liaohe River basin)	≥50-year-return extraordinary flood
2016	Taihu Lake basin	Basin-wide extremely catastrophic flood recorded as the second highest water level in the history
	Yangtze River basin	The most extraordinary basin-wide flood since 1998

2.1.1 Assurance of Lake and River Safety

The accumulated direct economic benefit obtained from flood control in China since 1949 amounts to 4.93 trillion yuan; the reduction of flood-inundated arable land areas totals 191 million hm^2, a yearly average reduction of 2.81 million hm^2; and reduction of grain loss due to flood control stands at 753 million metric tons, a yearly average of 11.07 million metric tons. In recent years, China has managed to handle several extraordinary floods, and successfully combated super typhoons, strong typhoons, tropical storms and severe local flash-floods.

2.1.2 Assurance of Water Supply Safety

From the 1990s onward, irrigation of drought-affected land extends to an area of 27.43 million hm^2 per annum on average, which reduces grain loss by 37.59 million metric tons, and enables 21.32 million people and 15.45 million livestock to have temporary access to drinking water. On multiple occasions, the Chinese government diverted water from the Yellow River to the City of Tianjin, and replenished fresh water to suppress salt water intrusion in the Pearl River basin, which has ensured the safety of water supply and ecosystems in Tianjin and cites on the Pearl River Delta including Guangzhou, Zhuhai, Zhongshan and Macau Special Administrative Region.

2.1.3 Assurance of Life Safety

With the enhanced management of society, and improvement of the flood control system, the death toll caused by flood disasters has decreased continuously year by year. In the 1950s, the number of deaths caused by flood hazards in China was 8,976 per annum while the corresponding figures in the 1980s and 1990s stood at 4,338 and 3,744 respectively. For the first 10 years of the 21st century, the death toll has been reduced to 1,582. Between 2011-2018, the death toll was reduced to 495. The death toll in 2018 was 187. Since 2010, in view of the fact that flash-floods are the main cause of casualties caused by floods, China has vigorously implemented the construction of flash-flood prevention and control projects. It has basically established a flash-flood disaster monitoring and pre-warning system and a mass monitoring and prevention system, which have played an important role. The average annual death toll from flash-floods in the first ten years of the 21st century was 1,079. From 2011 to 2018 after the implementation of the project, the average annual death toll from flash-floods dropped to 351.

2.2 Challenges

Presently, China encounters some problems and deficiencies in flood control and drought defense, including a lack of strategic water conservancy projects on some major rivers, shortage of flood control standards for some major tributaries, safety concerns of dams during the flood season, and difficulty in preventing and controlling floods in medium and small rivers, as well as mountain areas. The nation also encounters intensifying problems such as an under-developed allocation contingency plan for water conservancy projects, while the modern allocation platform is yet to be completed. On the other hand, hydrologic monitoring can't fully cover all the flood-control-functioning rivers. China still suffers from a weak capability of forecasting sudden regional heavy rainstorms and an unsystematic drought monitoring station network. Some major issues related to disaster prevention and control need our serious consideration in rural and urban areas, such as illegal appropriation of flood diversion rivers, floodplains and flood basins, waste of water resources, as well as disruption to water eco-systems. Besides, the public awareness of flood control, drought defense, risk prevention and emergency response is still yet to improve.

3. Strategies and Measures for Future Development

3.1 Strategies for Future Development

In 2016, President Xi Jinping proposed major principles of disaster prevention, control and relief. That is, we should put disaster prevention to the fore and employ an integrated approach of regular disaster control with emergent disaster relief. Meanwhile, we should strive to shift our focus from disaster relief to disaster prevention, shift our preparation for single disaster control to multiple disaster control, and shift our target from reducing disaster losses to reducing disaster risks. These principles offer a framework for the work of flood prevention and drought defense in the new era. Based on the shift of principal contradictions in China's flood control, the Ministry of Water resources stipulates guidelines for the reform and development of water conservancy projects. The guidelines should "Accelerating the construction of water conservancy projects and strengthen the supervision of water conservancy industry". With regard to flood control and drought defense measures, prevention will be a priority. We will focus on the scientific control of water conservancy projects and refine the administration of the industry for disaster defense. Meanwhile, our strategies lay emphasis on disaster monitoring and forecasting, disaster relief technology and water conservancy projects control. We will strive to handle the difficulties of mountain torrent defense and reservoir safety during flood season. To that end, it is necessary to conduct comprehensive examination and rectification of potential risks, revision and rehearsal of contingency plans, construction of the Watershed-Area Joint Disaster Control System of the Water Conservancy Projects, together with reservoir water level monitoring and supervision during flood season.

3.2 Main Measures for the Future Development

3.2.1 Secure Flood Prevention and Water Supply

We will closely monitor precipitation, flood and drought development and intensify consultation and investigation so as to release timely pre-warning and offer guidance for the scientific disaster-control of water conservancy projects at the watershed or local levels. It is also our duty to provide technological support for disaster relief, secure the life and property safety of the public, as well as the water supply in urban and rural areas.

3.2.2 Advance Capacity of Monitoring, Forecasting and Early-warning

We will strive to advance the capacity of river flood forecasting, monitoring and early-warning of medium and small rivers and medium and small-sized reservoirs, as well as emergency monitoring. Also on our agenda are intensifying mountain torrent prevention and control, and the construction of a flood forecasting and early-warning system in rural areas. We will improve the early-warning notification mechanism, in an attempt to guarantee prompt notice and full coverage, offering information support for risk avoidance and relocation.

3.2.3 Improve the Disaster Control Capability of Water Conservancy Projects

We will revise the regular and emergent water-transfer control schemes for major rivers, lakes and major water conservancy projects during flood or drought seasons. It is necessary to improve joint control of water conservancy projects, so as to make full use of the water conservancy network. To that end, we will conduct investigations on the multi-functional and coordinated joint control of the projects, apart from the main target of flood control. Furthermore, we will work on the watershed-area joint disaster control system of the water conservancy projects.

3.2.4 Intensify Supervision on the Control and Use of Water Conservancy Projects

In an attempt to regulate reservoir control and use, and eliminate illegitimate operations above the maximum water level during flood season, we will strictly implement the administrative regulations stipulated by the Ministry of Water Resources. These include the Regulations on *Supervision of Water Levels during Flood Season*, and the *Regulations on Inspections of the Disaster Control and Use of Water Conservancy Projects*. These measures will be accompanied with intensified supervision and accountability inspections for flood control and drought defense, so as to guarantee that the projects are under control in a legitimate, legal and scientific manner.

3.2.5 Facilitate Public Awareness of Disaster Defense

We will work on the evaluation of floods, and move on to the compilation and application of flood-risk charts. Also, we will carry out massive educational campaigns about flood and drought prevention and drought defense, so as to increase public awareness of disaster prevention. We will attach equal importance to the study and application of modern equipment and disaster relief technologies.

4. International Cooperation and Exchanges

China attaches great importance to international exchanges and cooperation in the fields of flood control, drought relief and disaster reduction. In recent years, five consecutive Sino-Swiss seminars on Flood Control and Disaster Reduction have been jointly held by China and Switzerland. China also hosted a technical workshop on the project of Urban Flood Risk Management sponsored by the World Bank, two training courses of the World Bank, and the Fifth Training Course on Regional Flood Risk Management. In addition, China has taken an active role in international exchanges, seminars and meetings. In October

2011, at the invitation of the Thai Government, China dispatched a team of flood control experts to Thailand for the provision of consulting services on flood control, disaster rescue and relief work. In June 2012, China sent another team of 23 experts to Thailand who advised on flood control and submitted the Consulting Report on Flood Control in the Chao Phraya River Basin in Thailand. In 2013, China and Russia cooperated to combat the flood in the Heilongjiang River basin. In 2014, the Ministry of Water Resources of China and the Emergency Department of the Russian Federation signed the MOU on flood control and hosted the first meeting of the Sino-Russian Working Group on Flood Control Cooperation. In 2015, at the request of the Myanmar government, China sent a team of experts to Myanmar for the provision of consulting services on emergency response and flood control techniques.The experts advised the

The Dyke along the Qiantangjiang River

Myanmar government on the emergency restoration of water-damaged projects as well as flood control and disaster mitigation planning. In 2016, China urgently dispatched cascade power station of Lancang River to increase water release to ease severe drought plaguing in Vietnam. In 2018, the Yarlung Zangbo River was threatened by two dammed lakes. The Indian was informed of the situation without delay. In addition, China also actively carries out joint international research projects. In cooperation with the World Bank, Asian Development Bank and International Council for Science, China has completed multiple projects such as research on major technical problems in flood control and disaster relief, regional consulting meeting on poverty and floods, research on flood management strategy, and research on drought management strategy in China, etc.

Chapter 10
Small Hydropower

Written by: Zhang Xuejin
Translated by: Shi Xianglin

1. Overview

Small hydropower refers to hydropower stations with an installed capacity of 50,000 kW or below and their matching grids. China ranks first worldwide in terms of small hydropower resources. The technically exploitable potential of small hydropower in mainland China is 128 million kW. The annual power generation is over 535 billion kW·h and small hydropower stations are distributed in 1,700 counties (cities) across the country in 30 provinces (autonomous regions and municipalities directly under the central government). At the end of 2018, there were over 46,515 small hydropower stations in operation with a total installed capacity of 80.4353 million kW, and a total annual power generation of 234.561 billion kW·h, which accounted for 22.83% and 19.02% of national hydropower installed capacity and annual power generation.

Installed Capacity Structure Diagram of Water and Electricity by Year-end 2018 in Rural Areas

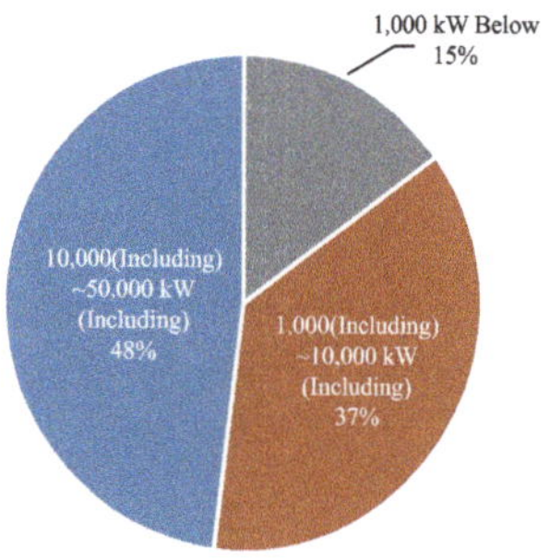

Installed Capacity Structure Diagram of National Power Supply in 2018

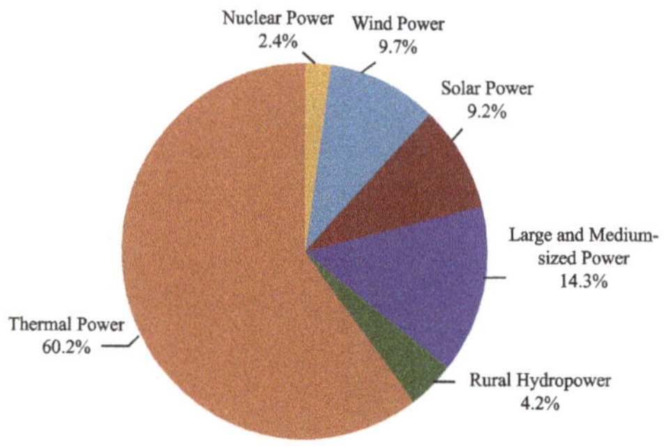

2. Major Achievements

In the process of developing small hydropower, China has built up a technical standard system that covers multiple aspects of small hydropower including planning and design, construction and installation, operation and maintenance, safety evaluation, and electro-mechanical equipment and material testing. As of 2018, there are currently 57 valid standards, including 16 national standards and 41 water conservancy industry standards.

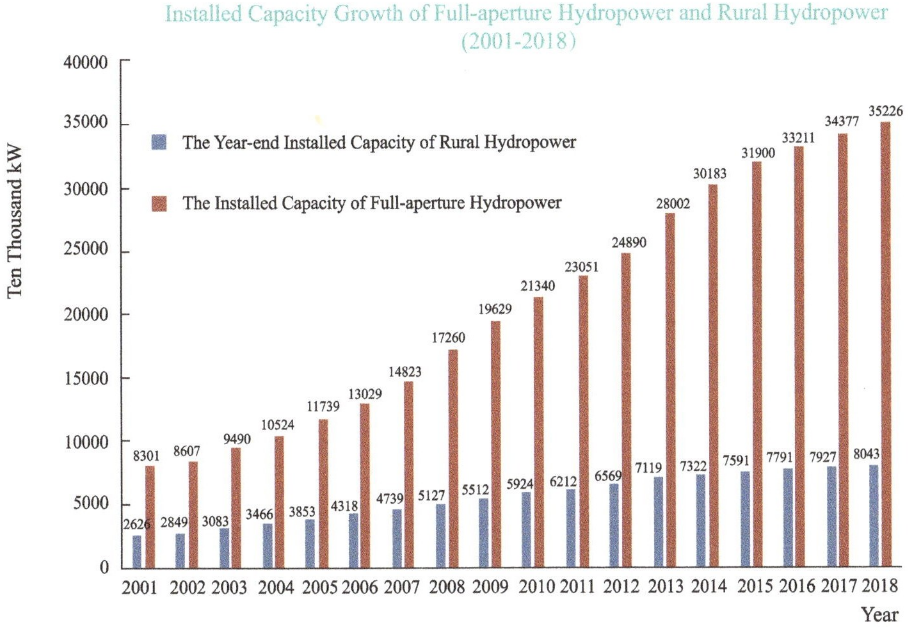

2.1 Small Hydropower, Big Role

2.1.1 Solve the Problem of Electricity Consumption for More Than 300 Million People Without Electricity

In 1983, the Chinese government initiated a campaign to enable its rural areas to get access to electricity through the development of hydropower. 653 counties accomplished preliminary hydro-powered rural electrification between the"7th Five-Year Plan"and the "9th Five-Year Plan"period. 621 counties accomplished hydro-powered rural electrification with higher standards between the "10th Five-Year Plan" and the "11th Five-Year Plan", and 316 more counties implemented hydro-powered rural electrification programs during the 12th Five-Year Plan period. The ratio of rural households with access to electricity grew from less than 40% in 1980 to 99.8% in 2015, along with a marked improvement in the quality and reliability of power supply, which basically resolved the problem of difficult access to electricity in mountainous rural areas.

2.1.2 Drive Forward Social and Economic Development in Rural Areas

The development pace of the counties enjoying access to preliminary hydro-

Jingrun Small Hydropower Station in Jingning County, Zhejiang Province

powered rural electrification between the 7th Five-Year Plan and 9th Five-Year Plan periods was remarkably higher than the national average. Those counties have basically achieved the targets of doubling their Gross Domestic Product (GDP), fiscal revenue, per capita net income of farmers, and per capita power consumption every 5 years, and quadrupling those figures every decade. In counties with access to hydro-powered rural electrification during the 10th-11th Five-Year Plan periods, their power consumption per capita and power supply reliability increased remarkably. Counties with newly gained access to hydro-powered rural electrification during the 12th Five-Year Plan period increased the installed capacity of the rural hydropower by over 5 million kW, which accounted for over 50% of the total national incremental installed capacity of rural hydropower.

2.1.3 Improve Farmers' Living Standards and Production Conditions

In many areas, the development of small hydropower plants helps improve local infrastructures, and promotes the development of public welfare. Isolated mountain villages have gained access to electricity, water supply and roads, and renovated their kitchens. Science and technology have been disseminated in rural areas. All these have brought enormous improvements to the spiritual outlook of rural residents in mountainous areas. In many villages with hydropower, the revenue from it helps fund culture, education, communications, medical treatment, social security and other public-good undertakings for the whole village. Seniors at and over the age of 60 are eligible for a "pension subsidy".

2.1.4 Contribute to Energy Conservation and Emission Reduction

In 2018, China's small hydropower generated more than 234.561 billion kW·h, equivalent to saving about 70 million tons of standard coal and reducing carbon dioxide emissions by nearly 100 million tons. The ecological protection project of replacing fuel with small hydropower implemented from 2003 to 2015 has solved the problem of 3.2 million farmers' living fuel needs and reduced the annual consumption of fuelwood by 5.7 million m^3. Forest protection areas were 733,300 hm^2. During the 12th Five-Year Plan period, more than 4,300 small and backward hydropower stations were upgraded and renovated.The consolidated and increased power generating capacity was equivalent to 11 million-plus tons of standard coal annually, reducing carbon dioxide emissions by 28 million tons, sulfur dioxide emissions by 248,000 tons and other harmful gases and dust.

Small Hydropower Station in Chaling County, Hunan Province

2.1.5 Safeguard Emergency Power Supply

When South China suffered freezing rain and snow storms in the beginning of 2008, small hydropower supplied starting power to a number of large power plants, and served as a secure source of electricity to over 200 counties and 2,000 townships during the Spring Festival for their power consumption. In addition, it provided power supply when the main power grids were in restoration. Small hydropower also helped to keep the major railway lines including Beijing-Kowloon, Yingtan-Xiamen, Chongqing-Huaihua, and Hunan-Guizhou uninterrupted. After the Wenchuan earthquake in Sichuan Province and Yushu earthquake in Qinghai Province, people in quake-stricken areas were able to quickly receive power supply in many counties and areas by restoring their small hydropower stations.

2.2 Strong Supervision, Green Development of Small Hydropower with a New Pace

2.2.1 Carry out Investigation Evaluation of Hydropower Resources and

Compilation of a Hydropower Resources Development Plan in Small and Medium rivers

In 2006, an investigation and evaluation of rural hydropower resources was initially carried out in China. The relationship of more than 16,500 rivers in the river basin and water system was comprehensively and systematically sorted

Improved Eco-environment Thanks to Small Hydro Replacing Wood Fuel Project

out. It was found that the technical exploitable amount of rural hydropower resources was 128 million kW, and the location distribution was basically the same as that of the population in the vast impoverished mountainous areas, ethnic minority areas and old revolutionary base areas. In 2012, the Ministry of Water Resources organized the revision of the hydropower resources development plan for more than 3,200 small and medium rivers in the country and optimized the distribution of rural hydropower

Shimenqiao Small Hydropower Station in Guangxi Zhuang Autonomous Region

resources in accordance with the principle of improving river ecology. This achieved good results.

2.2.2 Reinforce Rural Hydropower Safety Supervision and Ensure People's Lives and Properties

From 2011 to 2013, more than 46,000 rural hydropower stations in China were organized to appoint safety supervision bodies and production safety responsibility bodies station by station. Since the formation of production safety standardization was promoted in 2013, more than 2,500 power stations with production safety standardization have been built. The safety supervision system for rural hydropower stations has been continuously improved. Also, the government reinforced production safety supervision and inspection, intensified production safety education and training, and improved the safety level. Since 2006, the number of accidents and deaths in rural hydropower production have been gradually decreasing, and zero reports of personal casualty accidents occurred for the first time in 2015.

2.2.3 Implement Green Transformation of Small Hydropower and Restore River Ecology

In order to promote the restoration of river ecology, since 2011, the Ministry of Finance and the Ministry of Water Resources have allocated a special fund of 13.1 billion Yuan for renewable energy. Taking a river as a unit, the backward hydropower stations were upgraded and renovated, resulting in a consolidation and increase of installed generating capacity. The generating capacity has exceeded 12 million kW and 45 billion kW·h. Nearly 1000 rivers have been harnessed and over 1,800 km of dehydration-reducing rivers have been restored. All of these have basically eliminated river dehydration, disconnection and other watershed ecological problems caused by hydropower stations. The ecological restoration of rivers has achieved remarkable results.

2.2.4 Reinforce the System Design and Promote Green Development for Transformation and Upgrading of Small Hydropower

In 2006, the *Administrative Measures for Environmental Protection of Rural Hydropower Construction Projects* and the *Code for Environmental Impact*

Assessment of River Basin Planning were issued. In 2016, the Ministry of Water Resources issued the *Guiding Opinions on Promoting the Development of Green Small Hydropower* and the *Evaluation Criteria for Green Small Hydropower*. In 2019, the Ministry of Water Resources issued the *Notice of the Ministry of Ecological Environment on Strengthening the Supervision of Ecological Flow of Small Hydropower Stations in the Economic Belt of the Yangtze River*. 10 provinces and cities including Fujian, Zhejiang, Jiangxi, Henan, Hubei, Guangdong, Chongqing, Sichuan, Shaanxi and Gansu were guided to issue documents on the supervision and implementation of the ecological flow of hydropower stations. Fujian Province has issued the first *Measures for the Administration of Ecological Electricity Prices for Hydropower Stations (Trials)* in China and implemented the online electricity price reward and punishment policy depending on the ecological flow. Since 2017, a total of 165 small green hydropower stations have been built in 19 provinces across the country, which has played an exemplary role in promoting the improvement of ecology, benefiting people's livelihood and standardizing construction of small hydropower stations.

3. Work Goals and Main Tasks

3.1 Development Goals

During the 13th Five-Year Plan period, 5 million kW of installed capacity will be added. Green small hydropower construction will be carried out. Taking a river as a unit, the process of small hydropower with efficiency enhancement and expansion will be implemented. The ecological environment in 1,000 small and medium rivers will be restored. 2,000 old hydropower stations with efficiency enhancement and expansion will be completed. 1,000 hydropower stations will achieve ecological transformation. A new and improved 700,000 kW installed capacity of small rural hydropower stations for poverty alleviation will continuously help registered rural poor households. 3,000 hydropower stations with production safety standardization will be constructed. At the same time, the rural hydroelectric planning and standard system will be further perfected. The modernization level will be comprehensively improved. During the 14th Five-Year Plan period, the dehydration-reducing problem of rivers caused by rural hydropower stations will be fully solved. What's more, the government will increase efficiency of old hydropower stations, enhance safety guarantees, create a batch of green small hydropower stations, implement rural hydropower poverty-relief projects, and contribute to rural revitalization.

3.2 Main Tasks

3.2.1 With the Principal Line of Green Small Hydropower Construction, Promote Scientific Development

Great efforts will be made to build small hydropower stations, strengthen the planning of hydropower resource development in small and medium rivers, and optimize planning of hydropower resource development in river basins. The newly-built hydropower stations should follow construction procedures, and comply with the regulations on environmental protection and soil & water conservation, ensuring that rural hydropower development is harmonious with ecological environmental protection.

3.2.2 Restore River Ecology in the Process of Efficiency Improvement and Capacity Expansion of Rural Hydropower Stations

We should take improving efficiency, expanding capacity and transforming rural hydropower stations as an opportunity to implement all restoration measures of river ecology. We should enforce ecological regulation of river water and ensure ecological water flow when we try to improve efficiency of water resources, and also reinforce and increase the supply of renewable energy, and promote energy conservation and emission reduction.

3.2.3 Make Full Use of Small Rural Hydropower Plants to Alleviate Poverty

We should make the most of the advantages of small rural hydropower in poverty alleviation and establish a direct benefit mechanism for poor households while promoting scientific development of small hydropower stations in poverty-stricken areas. Besides, we should enforce small rural hydropower poverty-alleviation

Shilongba Hydropower Station in Kunming City of Yunnan Province, Constructed in 1911 (the First Hydropower Station in China)

projects, and constantly improve institutional mechanisms to make continuous contributions to poverty alleviation.

3.2.4 Strengthen Concurrent and Post-event Supervision to Ensure Production Safety

We should further strengthen the responsibility system for flood prevention and "duel responsibility" system for production safety, improve supervision to standardize production safety inspection and potential risk removal. At the meantime, we need to enhance standardization of rural hydropower stations to raise the level of production safety, and comprehensively improve hydropower production safety in rural areas, ensuring safety for the public and production.

4. International Cooperation and Exchanges

4.1 Establishment of International Organizations

In 1981, the Asia Pacific Small Hydropower Research & Training Center was established under the auspices of the UNDP and the then PRC Ministry of Foreign Trade and Economic Cooperation (MOFTEC). In 1992, following a joint initiative by international organizations including the UNDP, UNIDO and UNESCO, relevant countries and the Chinese government, the International Network on Small Hydropower (INSHP) was established in 1994 with a membership of over 60 countries. In addition, the International Center for Small Hydropower was established in 1999 with Hangzhou as its headquarters. At present, over 480 members from 80 countries and regions in the world have joined INSHP. Hitherto, the INSHP has established 3 regional sub-centers in Asia, Africa, and South America respectively, and 5 international small hydro demonstration bases in Zhangye of Gansu Province, Changsha of Hunan Province, Chenzhou of Hunan Province, Jinhua of Zhejiang Province and Nanjing of Jiangsu Province.

4.2 Construction of Platforms for International Small Hydropower Exchanges

Since 2005, the INSHP has successfully organized Hydropower for Today Forums 7 times, featuring various themes such as Asia-Africa Small Hydropower Cooperation, Developing & Managing Small Hydropower in Africa, Small Hydropower and Improvement of Living Standards, and Small Hydropower and Green Development. The forum held the Ministerial Seminar on Water Resources & SHP for Developing Countries and the Workshop on Water Resources Management & Planning for Developing Countries in Hangzhou in 2011 and 2015 respectively. In 2013, the International Center on Small Hydropower and UNIDO jointly launched the first English version of the World Small Hydropower Development Report 2013, and the second edition was released in 2016. The publication of SHP News, a special English journal on small hydropower and an INSHP Newsletter, the English journal of INSHP, further diversified such exchange platforms.

4.3 Engagement in International Cooperation and Technical Exchanges

Under the auspices of the United Nations and other international organizations, China has carried out international small hydropower cooperation programs in collaboration with the World Bank, the UNDP, the UNIDO, the Group of 77, the Global Environment Facility and other international organizations and international assistance agencies from developed countries. These programs were in the fields of rural energy, environmental protection and poverty reduction. They have helped promote and demonstrate small hydropower technologies, and enabled Chinese experiences, wisdom and technologies to benefit other developing countries. By engaging in inter-governmental technical cooperation programs, including the comprehensive development of small hydropower in the developing countries of Asia, Africa and Latin America, China-Africa Renewable Energy Cooperation and Sino-Pakistan Joint Research Center for Small Hydropower Technology, China has provided effective and practical assistance to developing countries in their use of small hydropower resources. The Lighting-up Africa Program has been developed. Small hydropower equipment developed and manufactured by China has been exported to more than 50 countries/regions, with some EU member states included. Services ranging from planning, designing and consulting to supply and installation of complete sets of equipment have been rendered to hundreds of small hydropower stations in developing countries.

The INSHP and Small Hydropower Research Training Center in the Asian-Pacific Region have provided small hydropower technical training, technical information exchange and scientific-technological cooperation for developing countries. 109 training sessions of small-sized hydropower technology have been carried out for countries participating in the Belt and Road Initiative. The number of trainees is over 2,500. Also, Vietnam, Guyana and Cuba or other Asian, African and Latin American countries have been assisted to construct a batch of small hydropower station projects. Planning, designing, consulting and complete equipment sets in hundreds of small hydropower station plans have been provided for more than 60 countries, including Mongolia, Rwanda, Brazil and Vanuatu.

In July 2016, Ban Ki-moon, Secretary General of the United Nations at the time, visited the INSHP to fully affirm the contributions of Chinese small hydropower stations in tackling climate change and developing small hydropower stations worldwide, as well as an important role in international exchanges.

On April 27, 2019, the International Standard Compilation of Small Hydropower, an important achievement of international cooperation under the Belt and Road Initiative, was included in the achievement lists of the 2nd International Cooperation Summit Forum of the Belt and Road.

In April 2019, E Jingping, Minister of the Ministry of Water Resources, Tian Shihong, Director of Standardization Administration of China, and Li Yong, Secretary-general of UNIDO together signed the MOU on the Development of International Standards for Small Hydropower

Chapter 11
Hydrologic Monitoring and Administration

Written by: Yang Jianqing
Translated by: Xiao Rui

1. Overview

Hydrology is a science that studies and analyzes the spatial and temporal distribution and changes of water. Based on various hydrologic data from the hydrologic station network, hydrologists analyze, predict and evaluate possible occurrences of floods or drought, as well as the spatial and temporal distribution and changes of water resources. Hydrology plays an essential role in the decision-making process of major social and economic affairs. These include flood control, drought defense and relief, development, the use and management of water resources, protection of the water eco-system, and the construction and operation of water conservancy projects. Hydrological affairs in China are administered by both the central government and provincial governments, characterized by a joint-administrative management mechanism. The Department of Hydrology, a part of the Ministry of Water Resources, administers all sorts of hydrologic affairs throughout the country. Bureaus of Hydrology in watershed areas, local governments and branches of the Department of Hydrology, are responsible for the hydrologic affairs of major rivers and their main tributaries as well as supervising the hydrology-related affairs of local subsidiaries.

China has a long history of hydrologic study, which could be traced back to 4,000 years ago. The legend of Da Yu Taming the Flood informs us that Chinese people knew the law that "water goes down to the low ground" thousands of years ago. Li Bing, a Chinese hydrologist that lived 2,200 years ago, set up a man-shaped stone scale to measure the water level at Dujiang Weir. During the Sui Dynasty (A.D. 581- 618), hydrologists used wood posts, steles as scales, and even carved lines on riverside rocks to measure water levels. In the subsequent 12 centuries from the Tang Dynasty (A.D. 618 - 907) to 1949, there were 72 records of the water-level going below the stone fish carving scale at the Baiheliang Ridge (White Crane Ridge) of the Yantze River. This indicates severe historical droughts and marks the longest field observation of water level in history. In modern times between the mid-19th century and 1949, China initiated systematic hydrologic study, including hydrologic observation, information transfer, document compilation, analysis and computation.

The past 70 years since 1949 have witnessed great progress of hydrology in China. China has achieved substantial accomplishments in hydrologic station networking, monitoring, forecasting, analyzing and computing, as well as water-resource evaluation, scientific research, specialisation, and international exchanges and cooperation. Since the Millennium, much more progress has been made in China due to larger investment and financial support from the central government, which has resulted in better hydrologic infrastructure, well designed and informative hydrologic station networks, modernized technologies and equipment. The great strides of hydrology are also characterized by a significantly improved capability of hydrologic forecasting and information services, accompanied by the establishment of a sound legal system, and a well-organized and specialised team striving for scientific research and innovation. The country has set up a relatively comprehensive hydrologic station network system, which plays a significant role in both hydrologic study and social and economic development.

1.1 Establish a Comprehensive Hydrologic Station Network System

Hydrologic stations are fundamental for hydrologic study and related work. Since 1949, China has carried out national hydrologic station network design four times so as to optimize the scientific design of the hydrologic network. In 1956, the Ministry of Water Resources drafted China's hydrologic network

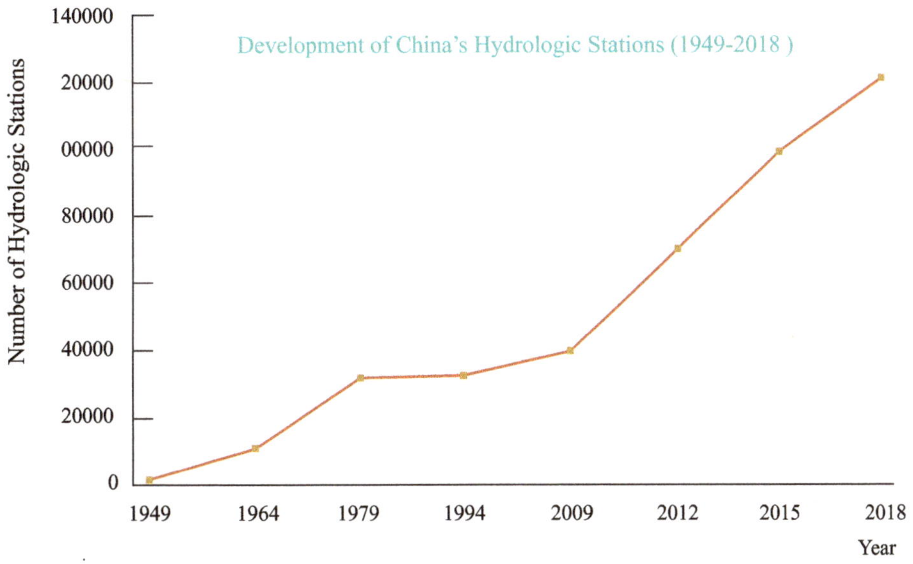

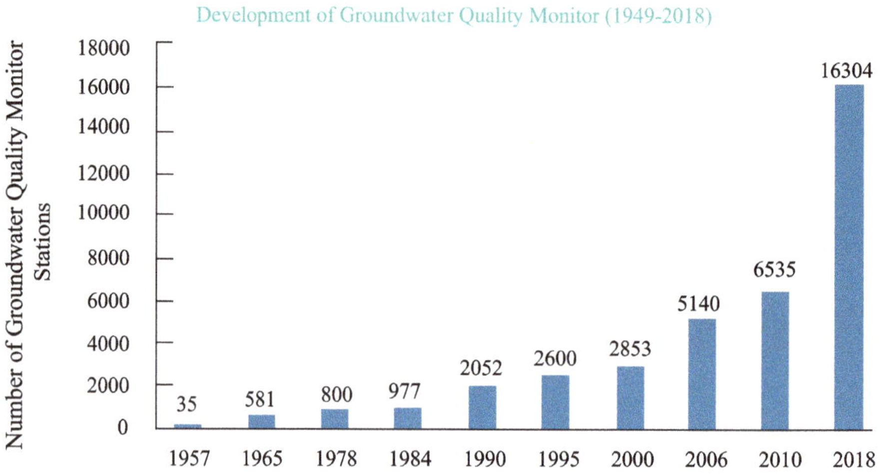

plan. Subsequently, the planning draft has been revised three times. In 2003, the Ministry of Water Resources released the *National Hydrologic Development Plan*. Since the reform and opening up policy, China has attached great importance to hydrologic work. The undertaking has been gaining momentum since the 18th Congress of the Communist Party of China (CPC). The central government has carried out many major hydrologic projects such as the Small and Medium River Hydrologic Monitoring System Project, the Inter-provincial River Hydrologic Station Network Project, the National Water Resources Monitoring Improvement Project and the National Underground Water Monitoring Project. These efforts have resulted in an increase of hydrologic stations multiple times and a huge expansion of monitoring coverage as a consequence. The total number of hydrologic stations in China was 121,000 at the end of 2018, up from 353 in 1949. Among other stations, there are 16,000 surface water quality monitoring stations and 9,000 ground water quality monitoring stations, which monitor the water quality of major places and watersheds in China, such as major rivers, lakes, reservoirs, drinking water sources,water diversion project hubs, inter-national (provincial) watersheds, ground water-monitoring wells, water function zones, etc. In the meantime, these stations offer comprehensive and timely services to the authorities regarding the changing quality of national water resources.

1.2 Massive Progress in Hydrologic Monitoring Capability

The Laboratory Technician with the Miyun Reservoir Monitoring Center, Beijing Water Environment Monitoring Center, Measures Water Temperature and Stabilizes Water Sample on the Spot.

Hydrologic measurement and forecast is technology-intensive and technology-dependent. In the past two decades, China's hydrologic measurement technology has evolved from manual observation to automated data collection and measurement. Hydrologic observations of water level and precipitation have become entirely automated measurement and reporting processes. Due to the fast development of modern technology, massive amounts of hydrologic instruments and equipment have been utilized extensively in hydrologic observation work. These instruments, such as the Accoustic Doppler Current Profiler (ADCP), electric wave current meters and ultrasonic sounders, are utilized multiple times more than ever before. Modern equipment, such as rain gauge radars, drones and remote control measurements, high performance gas chromatography mass spectrometers, atomic absorption spectrophotometers, flowing analyzers and biotoxicity analyzers, have found their extensive use in hydrologic work. Consequently, the accuracy and efficiency of water quality analysis has surged greatly, accompanied with an increase of automation and

Network and Wireless Transmission

Hydrologic Professionals are Measuring Flood Tides

information levels of hydrologic monitoring. In addition, with social and economic development, China has expanded its hydrologic monitoring coverage from major rivers down to all major small and medium rivers. Previously, the monitored items were mainly water levels, water flow and precipitation. Nowadays, the monitored items are more comprehensive, including evaporation, water quality, sediment, ice formation, soil moisture, ground water and the water ecological system, etc. Among other things, water-quality monitoring has developed from a single-dimension monitoring of chemicals to a comprehensive monitoring of water resources that includes surface water, ground water, wastewater, sediments, aquatic organism, and precipitation. We monitor nearly 200 physical, chemical and biological hydrologic indicators, which basically covers all major groundwater cycles, in an attempt to measure water contamination, and to provide comprehensive support to water resource utilization, eco-preservation and eco-restoration.

2. Major Achievements

2.1 Flood Control, Drought Defense, and Disaster Relief

Up to now, China has set up 11,000-plus hydrologic stations on 5,000-plus small and medium rivers for flood control. In addition, 1,700-plus hydrologic stations and major large reservoirs on the 170 large rivers throughout the country can offer flood forecasting reports, exceeding 90% of forecasting accuracy rate during critical periods. Over the country, 6 watershed authorities and 20 provinces (autonomous regions or municipalities directly under the central government) have put into effect local regulations on disaster alerts and information release, based on 700-plus alert indicators at major river sections. These measurements have contributed to promoting public awareness of disaster prevention and risk avoidance, and minimizing losses from disasters.

2.2 Administer Water Resources

The Ministry of Water Resources has basically set up an inter-provincial-section hydrologic station network of 53 rivers and monitoring system. These meet the demands for the water diversion of inter-provincial rivers, as well as the strictest supervision and examination standards of the water resources management system. The National Underground Water Monitoring Project, jointly constructed by the Ministry of Water Resources and the former Ministry of Land and Resources, is in pilot operation. The project set up 20,466 ground water monitoring stations throughout the country, examining the level, temperature, discharge, and quality of ground water over an area of 3.5 million km^2. The project has established a relatively comprehensive dynamic monitoring system of ground water in China, focusing on terraces such as large plains, basins, the Karst mountain areas, major water resources and overdraft areas.

2.3 Protect Water Eco-system

Hydrologic monitoring has developed from a single-dimension chemical analysis to a comprehensive monitoring of water resources that includes surface water, ground water, wastewater, sediments, aquatic organisms. We monitor nearly 200 physical, chemical and biological hydrologic indicators that basically cover the entire groundwater cycle, in an attempt to measure water contamination, and provide comprehensive support to water-resource utilization and water eco-preservation and restoration. Recently, under the guidelines of "heads of local government are responsible for all rivers in

Hydrologic Monitoring of the Closure Section of the Three Gorges Hydro Project

the region" (river chiefs and lake chiefs), local governments have attached importance to water quality monitoring and the observation of inter-regional river sections and major water system nodes. These measures guarantee the implementation of the "river and lake chief accountability" system.

2.4 Construct and Operate Water Conservancy Projects

To better serve construction and operation of water conservancy projects, the department has actively engaged in document compilation, and established special hydrologic, water-level and precipitation stations. Accurate computation, together with scientific analysis and evaluation, and precise fundamental data forecasts, plays a significant role in the planning, designing, constructing and operating processes of water conservancy projects.

2.5 Respond to Water-related Emergencies

The Department of Hydrology has initially set up an emergency treatment system that features quick response, advanced technology and adequate logistical support. The system facilitates our significant progress in handling emergency disasters. Using the system, the Ministry of Water Resources has offered prompt and timely monitoring information in the past years, with analysis and forecast reports on a series of emergencies, including Tangjiashan dammed lake during the Wenchuan earthquake, the Super-large Zhouqu mudslide, explosive contamination of the Songhuajiang River, and the Baige dammed lake of Jinshajiang River.

2.6 Serve Economic and Social Development

The departments have advanced open access of hydrologic documents and information. Government departments and the pubic could get access to speedy hydrologic information services through the hydrologic information system, subscriptions to official Wechat accounts, and various briefings, bulletins and special issues. The Department of Hydrology undertakes the National River and Lake Census, compiling knowledge on the basic information of rivers with a running area of over 50 km^2, and lakes that cover a permanent area of 1 km^2 throughout the country. The departments also provide monitoring and forecast services such as water level, flow speed and water temperature at waterfront tourist areas. In addition, during some major events, such as the Olympic Games and the World Exposition, the departments offer quality hydrologic support.

3. Major Work and Measures

3.1 Develop a Sound Hydrologic Legal System

In 2007, the State Council put into effect *Hydrological Regulations of the People's Republic of China*. The Regulations explicitly state the fundamentally public-welfare nature of hydrologic work. Meanwhile, the Regulations stipulate the hydrologic management system, and a series of other systems such as hydrologic development planning, station and network construction and management, information forecasts and releases, and environmentally-friendly monitoring instruments and equipment. In accordance with the Regulations, 26 provinces (autonomous regions and municipalities directly under the central government) have promulgated local hydrologic policies or regulations, which, together with the Regulations, constitute a sound hydrologic legal system.

3.2 Carry out hydrologic innovation incentives

Following the principles of innovation-oriented development, the departments have made prominent progress in technology standardization, research and development, and dissemination and application of research findings. In an attempt to encourage and promote hydrologic academic exchanges and innovation, the Ministry of Water Recources set up Liu Guangwen Hydrologic Technology Education Fund at Hohai University in October, 2010. The following March, the Ministry of Water Recources released the *National Hydrologic Science and Technology Development Plan*. It is noteworthy that China made an early step in hydrologic standardization issues. In 1955, China put into effect its first hydrologic technology standards - *Provisional Regulations on Hydrologic Measurement Methods*, which marked the embarkation of Chinese hydrologic standardization efforts. At present, China has set up 214 hydrologic technology standards, boasting a comprehensive hydrologic technology standard scheme that ranges from station and network management, monitoring technology and information processing to information forecasting, analysis and evaluation.

3.3 Organize hydrologic specialist teams

In China, there are now over 66,000 hydrologic professionals, compared with 756 in 1949. Recently, hydrologic departments have innovated methods of specialist team-building, developing a three-level specialist team-building scheme. This comprises hydrologic professionals at state and provincial level, hydrologic staff at municipal and county level, and hydrologic talents at the social level. 56 universities, colleges or institutes, including Hohai University, Wuhan University, and Tsinghua University, set up hydrology majors. They produce many qualified hydrologic graduates, which guarantees a healthy and consistent development of hydrology in China. Since 1992, the Ministry of Water Recources, the Ministry of Human Resources and Social Security, and the All-China Federation of Trade Unions have jointly sponsored the National Competition on Hydrologic Reconnaissance Skills on a regular basis. This

From March to May in 2016, Emergency Water Replenishment was Implemented to Mekong River through Jinghong Reservoir on the Lancang River.

competition has recognized many hydrologic reconnaissance craftsmen and skilled professionals. Also, the Ministry of Water Recources has organized 2,490 training sessions of all sorts since 2013, and a total number of 113,000 hydrologic technicians and managing staff have received the training.

4. International Cooperation and Exchanges

The hydrologic departments have made great efforts in international exchanges and cooperation. We consider it necessary to undertake our international obligations, and play our part through participating in major events of international organizations such as the IHP-UNESCO, the WMO, the IAHS, the IWRA, the TC, and the ISO. Meanwhile, we promote close international cooperation and exchanges with neighboring countries. We have conducted effective cooperation and exchanges with Kazakhstan, Russia, the DPRK, Mongolia, India, Bangladesh, Vietnam and the Mekong River Commission (MRC), in terms of trans-boundary river hydrologic data exchange, international flood reporting and hydrologic technology.

Chapter 12
Development of Water Science and Technology

Written by: Ru Nan, Guan Yuhui, Gu Jinyu, Zhang Jingguang, Meng Weichao, Liu Siyu, Cao Yan, Zhang Xiang, Zhang Xuejun, Qian Feng, Si Wei and Wang Bin
Translated by: Shi Xianglin

1.Major Achievements

Water science and technology has made noteworthy achievements in scientific research and technological development, as well as promoting and commercializing research findings. It serves the overall interests of the country, enhances its supporting role and ushers in future development centering on the main tasks of water conservancy.

1.1 Marked Achievements in Science and Technology Research in Water Conservancy

The scientific and technological innovation of water conservancy is actively promoted by focusing on the central work of water conservancy. "The Evolutionary Mechanism of the Water Cycle in River Basins and High-efficient Utilization of Water Resources", "Research and Application of Key Technologies for Dam Safety Assurance" and "Key Technologies and Application of Ecological Water-saving Irrigation Area Construction" won the first prize at the

Hangzhou Regional Center (Asia-Pacific) for Small Hydro Power Organized Trainees to Investigate the Three Gorges Project

Mayari Power Station in Cuba

Yaniz Hydropower Station in Turkey

National Scientific and Technological Progress Awards in 2014, 2015 and 2016 for three consecutive years. "The Three Gorges project" won the special prize at the National Scientific and Technological Progress Awards in 2019.

China's research on engineering sediment is taking a leading position worldwide represented by the research and practice of the sediment regulation system for large rivers and reservoir sediment reduction technology. China's theoretical sediment research keeps an advanced position internationally, represented by the theory of non-equilibrium sediment transport of non-uniform suspended load.

Hydrologic monitoring, pre-warning and forecasting techniques are among the world's most advanced. The accuracy of pre-warning and forecasting has been continuously improved. The prediction period has been constantly lengthened. The above progress strongly supports the remarkable improvement of flood and drought disaster prevention capabilities in China. In 2018, because of timely and accurate monitoring and forecasting by local hydrologic departments, the reservoir was scientifically dispatched to intercept floods and led to peak clipping and peak shifting, which effectively dealt with the great floods and hence made positive contributions to ensuring people's lives and maintaining social stability.

"Theory and Practice of Water and Sediment Regulation of the Yellow River" Won the First Class Award for National Science and Technology Progress in 2010

China leads the world in Dam construction technology. The "multi-level leap" with 100 m-level high dams, 200 m-level high dams and 300 m-level high dams has been realized. To be specific, the highest arch dam and rock-fill concrete dam in the world have been constructed, so that China owns the most 200 m-plus level high dams around the world. The Three Gorges Project, praised as "the pillars of a great power" by President Xi Jinping is the pivotal project with the most comprehensive functions of water conservancy and hydropower in the world. The construction technology of the smart dam has gained a breakthrough. Dozens of dams have obtained the outstanding achievement award of the International Fidic Engineering Project and International Dam Commission Milestone Project Awards.

The design and manufacturing level of giant hydro-generating units is also at a world-leading position. The single-unit capacity of the Baihetan hydropower project is up to 1 million kW, which is the largest single-unit capacity of a hydropower station in the world, and has full intellectual property rights.

Water resources allocation and high-efficient utilization have reached an international leading level. The construction technology of water transfer projects has been developed to an advanced ranking worldwide. The South-to-North Water Diversion Project is the largest water transfer project in the world, with a population of over 100 million directly benefiting from the first phase of the East Central Line projects. The theoretical research and practice in water resource allocation has made great achievements to enhance scientific support capacity with a water safety guarantee in Chinese basins.

The high-efficient water-saving irrigation technique is dramatically improving. The effective irrigation area in Chinese farmlands was expanded to 74 million hm2 today from 16 million hm2 in 1949, ranking first in the world. Supported by the high-efficient water-saving irrigation technique, China has irrigated farmlands that account for 50% of the national agricultural acreage to produce over 75% of total grains and more than 90% of cash crops, realizing the historical transformation of agricultural product supply from long-term shortage to basic balance.

1.2 Better Achievements Resulting from Introducing and Promoting Overseas Water Technologies

Starting from the 9th Five-Year Plan period, China has introduced more than 800 advanced technologies, equipments and facilities suitable for China's water resources from nearly 30 countries and regions. the following fields are included: development, utilization and rational allocation of water resources, disaster prevention and relief, rural water conservancy, water and soil conservation, monitoring of water environment and ecological governance, as well as project construction and management. The introduction of advanced technologies has significantly elevated the overall level of science and technology in water conservancy in China, and boosted its continuous future development. Through the work of the national fund for transformation of agricultural scientific and technological achievements, the water technology demonstration project and the local science and technology promotion plan, about 1,000 advanced and practical water scientific and technological achievements have been successfully put into popularization and application. They have played an important role in various fields such as disaster prevention and mitigation, efficient agricultural water use, soil and water conservation, and water project construction.

1.3 Initial Shaping of the System for Innovation in Water Science and Technology along with its Infrastructure Platform

Up to now, China has established a system for innovation in science and technology at the three layers of the State, basin and local administrative regions. The national innovation bases are located in two research institutions, namely, the China Institute of Water Resources and Hydropower Research (IWRH) in the north and Nanjing Hydraulic Research Institute (NHRI) in the south. A river-basin innovation platform has been built with the scientific research strength of the Yangtze River and the Yellow River basin institutions as the main body; Local water research centers and water science and technology test stations have been built with local water research forces as the main body. The formation of an infrastructure platform for water science and technology has been strengthened. The water industry has built 2 national key laboratories, 2 national engineering & technology research centers, 10 ministerial key laboratories and 13 ministerial engineering & technology research centers, 6 ministerial field scientific observation and research stations. A number of key laboratories, engineering & technology research centers and scientific research and testing bases have been built around the country, which enhance the ability and level of solving water-related problems.

1.4 Ever Enlarging Talent Pool in Water Science and Technology

At present, China has a pool of over 350,000 professionals and technicians in water conservation science and technology, among whom 172,000 have medium or advanced qualifications. Under the Ministry of Water Resources, there are 4 public welfare research institutes, in which 70% of all researchers have master or doctorate degrees. A large number of young talents in science and technology are trained. 85% of all researchers participating in national, provincial or ministerial science and technology projects are young specialists.

Ecological Water System for the Olympic Park

2. Challenges

China has a large population but less water and suffers the most prominent and complicated water problems in the world. The uneven temporal and spatial distribution of water resources in China is not compatible with the distribution of population, farmland, energy and mineral resources. In China today, problems such as drought & water shortage, flooding & logging, water pollution, and soil erosion remain acute, whereas the conflict between socioeconomic development and the support capacity of water resources and the environment keeps intensifying. With the progress of urbanization, industrialization, the creation of an ecological civilization and the implementation of national development strategies, coupled with the impacts of climate change, China is facing more severe water problems.

With regards to new factors and problems in water conservation, China should further speed up innovation in water conservation science and technology in line with the new requirements of socioeconomic development. Greater importance will be attached to the following areas: (1) research on the impacts of global climate change and large-scale human activities on water resources in China and possible countermeasures; (2) research on the strategies and countermeasures to solve problems such as drought & water shortage, flooding & logging, water pollution and soil erosion; (3) integration and combination of engineering technology, natural science and social science in water conservation; (4) promotion, commercialization and application of research results in water conservation science and technology; and (5) application of new advanced technologies in water conservation, so as to strive for new technological breakthroughs and achievements in major areas and key points of water conservation work.

3. Development Strategies

For future development of water science and technology, the water sector will adhere to the development concepts of innovation, coordination, green, open and sharing, carry out in-depth implementation of innovation-driven development strategy, and uphold the principle of water management, i.e., "prioritizing water saving, spatial balance, systematic governance, and equal attention to conservation and development". To address the four major pressing

The aqueduct in Shaoshan Irrigation District of Hunan Province

water issues in China, i.e., shortage of water, soil erosion, flood and drought, and severe water pollution, we will strengthen top-level design, promote optimal allocation of scientific and technological resources, deepen the reform of the science and technology system and establish and improve innovation. We will also work hard to eliminate the significant technical barriers in the development of water conservancy, effectively strengthen the promotion and implementation of advanced and practical achievements, and provide strong scientific and technological support for the modernization of water management system and governance capacity.

Scientific and technological innovation is the engine and power of water conservancy. Facing the new era with new tasks and new requirements, we should insist on linking long-term goals with current tasks. This includes firmly controlling the advancement and defining the goals of scientific and technological innovation for hydraulic engineering to strive for realizing breakthroughs as soon as possible. Besides, we should stick to combining strategic research with basic research, closely combining the scientific and technological innovation of water conservancy with the implementation and promotion of new, high-quality development concepts. This should reinforce the research on water safety guarantees related to the national strategy. What's more, we should adhere to the unity of scientific research and application of achievements, intensify research and development of advanced and practical technologies and products, and strengthen their popularization and application. Finally, we should accelerate upgrades of the standard, form the perfect standardized work management system, technical standards system, and standards supervision system. This will promote the internationalization of standards, and comprehensively improve our basic standardization capabilities.

Water and Soil Conservation Garden in Longli of Guizhou Province

封禁治理

Chapter 13
The Three Gorges Comprehensive Water Concervancy Project on the Yangtze River

Written by: Xu Hao
Translated by: Zhong Shumin

1.Origin of the Project

The Yangtze River is the mother river of the Chinese nation, with a drainage area of 1.8 million km^2. The middle and lower reaches of the Yangtze River are rich in products and densely populated, which is the focus of China's economic and cultural development, but is frequently stricken by severe floods, which have always been a major concern. In nearly two thousand years since 185 AD, there have been 219 hazardous floods, with an average of about one in 10 years. Every disaster has caused great loss of life and severe damage to people's property and the economy. After the policy of reform and opening up, and based on a large number of surveys, plans, scientific research, design, etc., with suggestions put forward by various circles, experts from all walks of life in the country organized a three-year in-depth investigation since 1986. Finally, the normal water storage elevation of 175 meters was recommended, and the construction scheme "first-class development, one-time completion, staged water storage and continuous immigration" was adopted. On April 3, 1992, the Fifth Session of the Seventh National People's Congress deliberated and passed the "Resolution on the Construction of the Three Gorges Project on the Yangtze River."

2.Construction of the Project

The Three Gorges Project includes hub projects, power transmission and transformation projects and resettlement projects. The hub project consists of a river dam, a power generation building and navigation buildings. The power transmission and transformation project bears the important task of delivering 22.5 million kW units of power generated by the Three Gorges Power Station.

The immigration project involves cities in the two provinces of Hubei and Chongqing, with a total of 1.3103 million migrants relocated and resettled.

The Three Gorges Project has been huge in scale, complex in technology and difficult to construct. The construction of the Three Gorges Project has utilized a combination of domestic R&D and imported technologies, and overcome a series of worldwide problems through introduction, digestion, absorption and technological innovation. The project includes the world's largest volume of concrete (16.1 million m^3), the largest number of holes in a dam body (83), the largest discharge gravity for a dam (100,000 m^3), the largest hydropower station in the world with an installed capacity of 22.5 million kW and an inland river lock with the largest scale and the highest head (113 m). The project has also constructed a vertical ship lift with the largest height (113 m) and the largest weight (15500 t) in the world and an innovative 500,000 VAC and DC line erection and dispatch communication system. Through scientific research and independent innovation, it has solved a series of major technical issues such as river closures, ship lock deep rock excavation (175 m), dam concrete continuous pouring and temperature control, heavy metal structure installation, and the manufacture and installation of 700,000 kW giant hydroelectric generating units. The Three Gorges Project has set 112 world records, won 18 national prizes for scientific and technological progress, and 1053 patents for technological innovation and invention.

Migrant relocation and resettlement for the construction of the Three Gorges Project has been critical to its success. The State Council promulgated and implemented the *Regulations on Resettlement for the Construction of the Three Gorges Project of the Yangtze River*, carried out the policy of exploitative resettlement, scientifically worked out the resettlement plan, and insisted on resettlement according to law. Nineteen provinces (autonomous regions and municipalities directly under the Central Government), 10 large and medium-sized cities and more than 40 central units in China supported the Three Gorges Reservoir Area. A total of 0.1962 million people walked out of the Three Gorges reservoir area and settled in rural areas along the Yangtze River or coastal areas. Enterprises that caused serious pollution, had no market for their products, or suffered from insolvency were issued with bankruptcy or closure.

The Three Gorges Eco-environmental Protection Project not only reduces the adverse impacts of the project's construction on the local ecological environment, but also improves the environmental state and water quality of the reservoir area. The prevention and control of geological hazards in the Three Gorges Reservoir Area ensures the water storage safety of the Three Gorges Project and the safety of the relocated cities and towns. The Protection of the Cultural Relics of the Three Gorges Project has helped to identify and protect ancient culture during the development of the Three Gorges Project and preserved and inherited the splendid ancient culture of the Three Gorges area.

3.Benefits of the Project

The Three Gorges Project began to store water at 135 m water level in June 2003. In 2006, it was stored at 156 m. After 175 m experimental water storage began in 2008 and the normal water level of 175 m was reached in 2010, it had already passed the full storage cycle and brought its benefits into full play.

3.1 Remarkable Flood Control Benefits

Before the construction of the Three Gorges Project, flood control in the middle and lower reaches of the Yangtze River mainly relied on dikes, flood diversion and storage measures. The standard flood control of the main embankments of the Yangtze River was usually once every 10 to 30 years. If the above criteria were exceeded, measures such as flood diversion should be adopted. Hundreds of thousands of people went to the dikes to patrol and rescue during the flood season every year; tens of millions of people were seriously threatened when severe floods occurred. The Three Gorges Project has a flood control reservoir capacity of 22.15 billion m^3, which can directly control 95% of the flood volume of the Jingjiang river section in the most rigorous flood control situation. Standard flood control for the Jingjiang River section has been raised from "once in a decade" to "once in a hundred years". In 2010 and 2012, the Three Gorges Project withstood two tests that exceeded the maximum flood peak of 1998, ensuring flood control safety for the middle and lower reaches of the Yangtze River. During the flood season of 2016, the largest flood since 1999 hit the middle and lower reaches of the Yangtze River. Through scientific dispatch, the peak of the Three Gorges Project was successfully cut by 38%, effectively alleviating the flood control pressure in the middle and lower reaches of the Yangtze River. In addition, the implementation of small and medium-sized flood storages has expanded the benefits of flood control, improved the utilization rate of flood resources and enhanced the safety of the Yangtze River

waterways. By the end of 2018, the Three Gorges Project had been used 47 times for flood control, with a total flood storage capacity of 150.7 billion m^3 and therefore created a secure environment for the economic and social development in the middle and lower reaches of the Yangtze River.

3.2 Excellent Power Generation Efficiency

Since all the 32 700-MW units of the Three Gorges Power Plant were put into operation in 2012, the annual average power generation capacity has reached 92.98 billion kW·h. In 2018, the power generated was 101.62 billion kW·h, making it the largest hydropower station in the world that year. By the end of 2018, the Three Gorges Power Station had generated a total of 1,190.5 billion kW·h. It has effectively alleviated the power shortage in East China, Central China and Guangdong, and has become an important large-scale clean energy production base in China.

3.3 Prominent Shipping Benefits

After the completion of the Three Gorges Project, the navigation conditions of about 660 km from Yichang to Chongqing were greatly improved. The water depth of the waterway below Yichang has been increased by 0.6 ~ 1.0 m, and 10,000-ton fleets can reach Chongqing directly from Shanghai. Shipping costs are notably reduced and safety is significantly improved. In 2018, the throughput of the Three Gorges Lock reached 142 million tons, a new record which was eight times the maximum annual throughput before the construction of the Three Gorges Project. By the end of 2018, the freight volume through the Three Gorges Ship Lock had accumulated to more than 1.255 billion tons, greatly promoting the exchange of materials between the southwest hinterland and the coastal areas.

3.4 Striking Benefits of the Comprehensive Utilization of Water Resources

The Three Gorges Reservoir is an important strategic freshwater resource reserve in China, guaranteeing water supply protection for water-deficient areas along the Yangtze River and northern China. The Three Gorges Reservoir utilizes a huge regulation storage capacity to “accumulate and replenish dryness”. The annual discharge flow during the dry season is increased from less than 3000 m^3/s to over 6000 m^3/s, and the water supply is more than 20 billion m^3, which has effectively eased the seasonal tensions in industrial, agricultural, production and domestic water use in nearly 60 cities and more than 150 counties in the middle and lower reaches of the Yangtze River. By the end of 2018, the accumulated water supply downstream of the Three Gorges Reservoir was 1,950 days and 243.2 billion m^3 during the dry season.

3.5 Effective Protection of the Ecological Environment

The Three Gorges Project provides clean energy, greatly reduces carbon emissions, and contributes to the mitigation of the greenhouse effect. By the end of 2018, the cumulative power generation of the Three Gorges Power Station was equivalent to 381 million tons of standard coal combustion, corresponding to a reduction of 259 million tons of carbon emissions, 950 million tons of carbon dioxide, 11.43 million tons of sulfur dioxide and 5.71 million tons of nitrogen oxides. The Three Gorges Reservoir implements ecological dispatching to help the natural breeding of four major Chinese carps and the suppression of salt water tides in the Yangtze River estuary. From 2011 to 2018, the Three Gorges Reservoir carried out 11 ecological dispatches, utilizing “artificial flood peaks” to promote spawning and reproduction of the four major Chinese carps, which brought about visible effects on restoring fishery resources in the middle reaches of the Yangtze River.

Moreover, with the completion and operation of the Three Gorges Project, the pattern of joint dispatch and allocation of inter-basin water resources has been developed among the group of reservoirs in the upper reaches of the Yangtze River with the Three Gorges Reservoir as the backbone.

4. Major Focus and Main Measures

4.1 Strengthen the Operation and Management of the Three Gorges Project

According to the concept of "ecological civilization of harmonious co-existence between man and nature", the joint dispatch of reservoirs in the river basin and the flood control guarantee system of the Yangtze River are continuously improved with the Three Gorges Reservoir as the backbone. Through intensified systematic management, the role of the Three Gorges Project in the ecological restoration and protection of the Yangtze River is given full importance. The carrying capacity of the water environment and the aquatic ecology are improved through ecological regulation. The functions of the Three Gorges Project in green

power and green shipping have been enhanced, and the high-quality water supply capacity of the Three Gorges Reservoir has been increased.

4.2 Protect and Manage the Three Gorges Reservoir in Accordance with the Law

We must refine the legal system and integrate the operation management of the Three Gorges Reservoir into the rule of law; consolidate planning and guidance; protect the shoreline and storage capacity of the Three Gorges Reservoir according to regulations; practically strengthen the management of the riparian zone to ensure the safe operation of the Three Gorges Project; research and establish a more stringent management system and mechanism for the Three Gorges Reservoir; actively promote the construction of the Three Gorges National Park; deepen the ecological sustainability of the river basin; and facilitate the formation of a long-term system for the protection of the Three Gorges reservoir area.

4.3 Continuously Improve the Management for Operational Safety of the Three Gorges Project

The operation of the Three Gorges Project involves flood control, sand erosion, power output, lock passage, downstream water supply, ecological dispatch and other aspects. It is necessary to bring into line various functions and coordinate the interests of all parties to build a long-term system to strengthen supervision and inspection. This will include an improved emergency response capability and safety supervision to build and perfect a comprehensive safety monitoring system for the Three Gorges Project. Observation and study of sediment and changes in river-lake relationships should be carried out to ensure the safety of embankments. We must continue the ecological restoration of the main stream of the Yangtze River, major rivers and lakes to ensure production and domestic water needs; to strengthen waterway remediation and dredging, and to continuously improve navigation support capabilities.

4.4 Promote the Construction of a Harmonious and Stable New Reservoir Area

Continue to strengthen the follow-up work of the Three Gorges Project, focus on the stability and prosperity of migrants and water security, and promote the development of high-efficiency ecological agriculture and featured precision agriculture. We shall also work on the integration of arts and tourism and actively develop tourism, optimize industrial structure, enhance employability, promote high-quality development in the reservoir area, and further strengthen water ecological restoration and water environmental protection. In particular, it

is essential to strengthen the management of important tributary systems in the Three Gorges reservoir area and promote harmony between people and water. Great efforts should be expended to promote the implementation of the strategy of Rural Revitalization and the construction of small towns for the purpose of creating ecologically civilized communities, and promoting social harmony and stability in the Three Gorges Reservoir Area.

Chapter 14
The South-to-North Water Diversion Project

Written by: Yin Litao
Translated by: Xiao Rui

1.Background

In China, the watershed regions of the Yellow River, the Huaihe River and the Haihe River are examples of severe water shortage. There is a population of 440 million living in the region, making up nearly 35% of China's total population, and 35% of China's GDP. In contrast, the regional water resources account for only 7.2% of China's total. Furthermore, the average amount of water resources per capita of the watershed of the Haihe River, where Beijing and Tianjin are located, is 292 m^3, accounting for only 14.3% compared to the nation's average.

The South-to-North Water Diversion Project (hereinafter referred to as the Project) is a major strategic infrastructure, which aims to alleviate severe water shortage in northern China, optimize water resources allocation and improve the ecological environment. The Project was initiated in 1952. With adequate demonstrations throughout the past 50 years, the best development plan was chosen from among 50-plus plans. The plan proposes a project that transfers water from the south to the north with three routes: the eastern, central and

western routes. Once completed, the Project, connecting the Yantze River, Yellow River, Huaihe River, and Haihe River, could be an integral part of China's water system. This water system facilitates water resource allocation throughout the country according to the "South-to-North transfer, East-West Complementary Transfer strategy".

The Project adopts the basic principle of "a priority of water conservation, pollution control and environmental protection". The total designated volume of water transfer is 44.8 billion m^3. The construction of the Eastern Route is carried out in three phases, the Central Route two phases and the Western Route also three phases. Once completed, the Eastern Route could transfer 14.8 billion m^3 of water, the Central Route 13 billion m^3, and the Western Route 17 billion m^3.

2.Construction of the Project

On December 27, 2002, the first phase of both the Eastern and Central Routes

of the Project kicked off. They were completed on November 15, 2013 and December 12, 2014 respectively. The successive construction of the two Routes have been listed on the Thirteenth Five-Year-Plan for National Economic and Social Development of China, with the construction of the Western Route making good progress.

After construction of the first phase, the 1,467 km-long Eastern Route, has 34 large pump stations distributed at 13-level cascades, with a water pumping capacity of 8.8 billion m^3 of water annually. The Central Route extends 1,432 km, with a water pumping capacity of 9.5 billion m^3 annually. The first phase of construction accomplished earth-rock engineering by 1.6 billion m^3 (280 million m^3 and 1.32 billion m^3 for the Eastern and Central Routes respectively), and

1,238 Highway-Crossing Buildings

concrete engineering by 42.82 million m^3 (5.93 million m^3 and 36.89 million m^3 for the Eastern and Central Routes respectively).

The Project has been the world's longest water transfer project with the largest area and population benefited. It is an aggregate of massive projects, facing major technical challenges during design, construction and operation. Despite these challenges, the authorities provided all kinds of resources to boost technological advancement. The efforts have been rewarding, and a series of technical problems and challenges have been tackled, such as the integration of old and new concrete during the Danjiangkou dam heightening project, Cross-Yellow-River technology of the Central Route, canal slope treatment in the swelling soil zone, efficient construction technology of large aqueducts, water transfer technology at ice seasons, large-discharge, and the low-head pump station construction technology of the Eastern Route. These new technologies, machinery, and materials have played a significant role in facilitating the quality and safety of the Project during construction and operation, many of which have become technical standards, patented and awarded with national prizes.

During the first phase of the Eastern and Middle Routes Project, 435,600 people have been relocated to another residence. Following the human-first principle, land appropriation and household resettlement work the Project consists of a two-step incentive: compensation and funding in the first place, and then policy and financial support when people have been resettled. These measures were well-received so that the relocation work was accomplished in two years, twice as fast as scheduled, and people were cooperative. Consequently, those resettled are better-off in their new homes now.

Local governments of the Water-Source areas and watershed areas of the Eastern and Central Routes take firm and effective measures to deal with water pollution in the the Danjiankou Reservoir and upstream areas, and with pollution prevention and control along the Eastern Route and heavily-polluted areas. These measures, together with the improvement of the eco-system along the main canals of the Project, secure a quality water-supply.

Protection of cultural and historical relics is another integral part of the Project, as it extends through many historically and culturally renowned areas. Protection and preservation of relics is the first and foremost task. To that end, the Routes design was revised multiple times. The Project has obtained a win-win achievement, that is, protection and preservation work of relics has been carried out along with Project construction.

3.Benefits of the Project

Since the completion of the first phase of the Eastern and Central Routes, the Project has been running smoothing, with water quality constantly above standard. The resettled people are generally better off. The investment in the Project is rewarding. By October 11, 2019, the Routes have transferred a total volume of 28.68 billion m^3 of water. The Eastern Route has transferred 3.91 billion m^3 of water to Shandong Province, the central Route 24.77 billion m^3.

3.1 Optimize Water Resources Allocation and Water Supply Capacity

The Project has basically solved the problem of water supply in northern China, improving water supply capacity and water quality of 40-plus large and medium-sized cities in the region. The transferred water is no longer a supplement source, but a major source of water supply in these cities, directly benefiting a total population of 120-plus million.

In addition, the cities along the Routes, receiving the transferred water from the South, have reduced over-withdrawal of underground water. Further, the transferred water helps to maintain underground water balance, which optimizes water resources allocation in the region. For instance, in Beijing, the Project provides 75% of water. The water supply safety index of the downtown area has increased from 1.0 to 1.2. In Tianjin, the Project offers water to 14 administrative districts. The city has secured its water supply from both the Project and the

Luan-River-to-Tianjin Project. In Henan province, 37 cities or counties have benefited from the Project. In Zhengzhou, 80% of tap water in the downtown area is from the South. Even more, 100% of downtown water is from the Project in such cities as Hebi, Xuchang, Luohe and Pingdingshan. In Hebei province, 80 cities or counties have received the transferred water. In Jiangsu province, the Project has secured a water supply for irrigating farmland covering over 3 million hm^2 in 50 districts and counties. The Eastern and Central Routes converge in Shandong province. The T-shaped Route intersection plays a significant role in the economic and social development of the province.

3.2 Improve Water Quality and Consequently Citizens' Livelihood

The Project offers a constant supply of quality water. Quality of the transferred water from the Eastern Route has been kept above Class 3 of the National Environment Quality Standard for Surface Water, and that from the Central Route, above Class 2. The Project has greatly improved water quality for those citizens in the North. Furthermore, in Beijing, the water hardness has substantially dropped from 380 mg/L to 130 mg/L.

3.3 Promote Economic and Social Development and Facilitate the Implementation of Key National Strategies

The watershed region of the Yellow River, the Huaihe River and the Haihe River is significant to the plans for the national economy. The Project has secured water supply for the successful implementation of key national strategies, such as the coordinated development of Beijing, Tianjing and Hebei province, and the development of the Xiongan New Area. According to the statistics of average water consumption per RMB 10,000 GDP output, the Project has supplied water for the development of 4 trillion Yuan in northern China.

3.4 Improve Eco-environment and Generate substantial Eco-benefits

By October 11, 2019, the Eastern and Central Routes have transferred 2.69 billion m^3 water. Rivers in northern China, including Baiyangdian, have seen a

51 Railway-Crossing Projects

significant rise in water level. As a result, the eco-system and water quality of the rivers has been improved, together with a rise of the underground water table. The Project has brought with it massive eco-benefits.

In 2017 and 2018, for two straight years, the Central Route supplied a total amount of 1.16 billion m^3 water to some rivers in northern China, discharged from the Danjiangkou Reservoir during the flood season. Since September 2018, the Ministry of Water Resources, together with Hebei province, has conducted joint efforts to comprehensively handle the problem of the over-withdrawal of underground water in northern China. As a key measure, the Central Route has supplied 930 million m^3 water to the Hutuo River, the Fuyang River and the Nanjuma River, with the largest river covering an area of 46 km^2. From April to June 2019, we initiated a pilot program -- emergent water transfer through the north extension of the first phase of the Eastern Route. This program transferred 37.39 million m^3 water to Hebei province, and 19.78 million m^3 water to Tianjin.

The pilot program was successful. The underground water resources in northern China have been conserved and restored, with a constant rise of underground water level, especially in the peripheral districts of some major water-depleted areas. It is noteworthy that the underground water level of the plain area of Beijing has been on the rise for three consecutive years since 2016. The water storage has increased by 1.39 billion m^3, and the water level has increased by 2.72 m. The Project plays a pivotal role in supporting and facilitating economic, social and ecological development of both the water-transferring and water-transferred areas. Meanwhile, the Project is generating massive social, economic and ecological benefits. It is becoming more strategically significant in national affairs like the scientific arrangement of water transfer, accurate and safe water supply, and standardized water allocation. The Project has been a model of optimizing water resource distribution by means of water transfer.

27 Aqueduct Projects

102 Inverted Siphon Projects

902-km Canal Projects

This book is the result of a co-publication agreement between China Water & Power Press (China) and Paths International Ltd (UK)

Title: China Water
Edited by Ministry of Water Resources of the People's Republic of China
ISBN: 978-1-84464-674-6
Ebook ISBN: 978-1-84464-675-3

Copyright © 2021 by Paths International Ltd, U.K. and by China Water & Power Press, China

All rights reserved. No part of this publication may be reproduced, translated, stored in a retrieval system, or transmitted in any form or by any means, electronic, mechanical, photocopying or otherwise, without the prior permission of the publisher.

The copyright to this title is owned by China Water & Power Press, China. This book is made available internationally through an exclusive arrangement with Paths International Ltd of the United Kingdom and is permitted for sale outside China.

Paths International Ltd
www.pathsinternational.com
Published in the United Kingdom

CPSIA information can be obtained
at www.ICGtesting.com
Printed in the USA
LVHW070425211121
703940LV00001B/4

9 781844 646746